PRAISE FOR MONK YUN ROU'S BOOKS

Turtle Planet

"Turtles have been a part of Earth's natural balance for hundreds of millions of years. Now, human greed and indifference bring them to the very brink of extinction. In this passionate, shining work, Yun Rou champions their cause and indicts our self-destructive relationship with Mother Earth."

—William Holmstrom, Wildlife Conservation Society

"This beautiful, imaginative, and important work reminds us that turtles, unchanged for 200 million years, have been a cornerstone of folklore and religion since before recorded history. If we can come to see the threats facing them today, as Yun Rou has done here, then we can begin to repair what we have done."

—Anthony Pierlioni, Vice President and Senior Director, theTurtleRoom

A Cure for Gravity

"[A] charming tale.... There's a bravura innocence at the heart of this offbeat novel."

—*Publisher's Weekly*

"A touching ghost story that eludes easy comparison to any other book. An amazing, rewarding voyage.... No need to imitate other writers; Rosenfeld is a true original."

—*Booklist*

"*A Cure for Gravity* roars along at the pace of an open-throttled motorcycle."

—*The Tribune*, South Bend, IN

"With *A Cure for Gravity*, Mr. Rosenfeld inspires the deepest emotion one writer can feel about another: envy."

—Larry Gelbart, creator of *M*A*S*H*, *Tootsie*, *A Funny Thing Happened on the Way to the Forum*, and more.

"A novel of surprising imagination and stylistic daring....*A Cure for Gravity* rises to near greatness as a piece of home-grown Magical Realism. Touching, scary, hilarious."

—Knight Ridder News Service

"This book is like reading a story and listening to music at the same time. A page-turner with rhythm, and a most unusual narrative voice. I loved the characters, the views of an America I haven't seen, the unexpected twists and turns. A wonderful book."

—Jack Paar, former host of *The Tonight Show*

"Arthur Rosenfeld's *A Cure for Gravity* is a noir mystery, a supernatural thriller, a crime caper novel, a love story, and an American road-trip adventure—all seamlessly woven into one moving, magical book. If the ghosts of Jack Kerouac and Jim Thompson could collaborate with Alice Hoffman, this is the story they might write.... This novel twists, spins, and rages like an Oklahoma tornado, and it'll fling you up into the cruel sky before bringing you back down to the good earth...safe, but shaken. Hell, it'll make you fly."

—Bradley Denton, author of *Blackburn* and *Lunatics*

Diamond Eye

"This is, to put it bluntly, one of the freshest, most enjoyable mysteries to come along in the last couple of years.... Any novel that features people with names like Seagrave Chunny, Phayle Tollard, and Twy Boatright is a novel that practically demands to be read.... The plot is delightfully twisty, turny, and, at times, surprisingly thought provoking."

—*Booklist*, **starred review**

"A great read in the noir tradition."

—**Cluesunlimited.com**

"*Diamond Eye* is a special delivery, no doubt about that. Refreshingly different. With its wit, warmth, and wonderfully wild cast, Rosenfeld dexterously blends cinematic scenes with intricate, often humorous personality studies in what may be this year's most promising detective series introduction. Hey, who knew that detective fiction could benefit from going postal?"

—*January Magazine*

"Rosenfeld writes a muscular prose that moves along at a brisk clip."

—*Florida Sun-Sentinel*

"Exploring cop-struggling-against-criminal-desire themes hauntingly reminiscent of Hammett's *Red Harvest*, Rosenfeld crafts a high-action suspense thriller with plenty of wry humor and cultural commentary."

—*Publisher's Weekly*

"Rosenfeld's likeable detective has a genuine disgust for the felons that he pursues and the determination necessary to bring them to justice."

—*The Dallas Morning News*

"Rosenfeld skillfully weaves a complex plot that defies solution until the very last pages. In the process, he creates an urbane, life-loving, self-effacing, and courageous character who should forever dispel the erroneous image of 'lowly' postal inspectors."

—*The Boca Raton/Delray Beach News*

"Rosenfeld keeps things moving smartly even before the nifty twist that ties his two plots together into a neat, grisly bow."

—*Kirkus Reviews*

The Cutting Season

"Arthur Rosenfeld's *The Cutting Season* is a marvelously entertaining blend of many different genres: medical thriller, psychological suspense, fantasy, martial arts adventure, romance, and crime drama, all neatly packaged into three hundred engrossing pages."

—**Mostlyfiction.com**

"Highly recommended, and not just for martial artists. This is a well written story that all will enjoy."

—**Larry Ketchersid, author of *Dusk Before Dawn***

"It takes a bold author to attempt the creation of a new category of popular fiction. That's the task crime novelist and tai chi master Arthur Rosenfeld set himself with his ninth novel, *The Cutting Season*."

—*Florida Sun-Sentinel*

"A gripping story...a page-turning mystery.... Rosenfeld's medical knowledge and martial-arts expertise reinforce an authority and clarity to the work.... That's storytelling!"

—**Walter Anderson, Chairman and CEO of *Parade Magazine***

"....lively, accurate, and beautiful writing...[this] secret world of blades...[is] brimming with romance, mysticism, and murder. It's the rare writer who can hold my interest so intensely."

—Dellana, Master Bladesmith

"A brain surgeon swordsman battles with...Russian mobsters, and his own reincarnations. This smart thriller sets a refreshing new standard for martial arts fiction."

—*Kung Fu Magazine*

"A writer who understands the deeper side of these sacred arts...a breath of fresh understanding."

—Stuart Charno, Shing-I Ch'uan Master

"A home run! We rate this book five hearts."

—*Heartland Review*

"...an intriguing premise as the hero rationalizes his vigilante justice... to do nothing would be amoral. Fascinating."

—*The Midwest Book Review*

"Remarkable!....a literary masterpiece...exceptionally well-paced and hard to put down...unique insights in the mysterious world of classical martial arts."

—Lawrence Kane, author of *Surviving Armed Assaults*

The Crocodile and the Crane

"Arthur Rosenfeld has done it again!"

—*Virginia Gazette*

"...a thriller of uncommon inventiveness. In the hand of the right filmmaker, *The Crocodile and the Crane* could be a terrific movie."

—*Florida Sun-Sentinel*

Tai Chi: The Perfect Exercise

"In *Tai Chi: The Perfect Exercise*, Arthur Rosenfeld draws from modern newsfeeds and a multitude of personal colorful anecdotes to illuminate this time-honored art. He brings a charmingly refreshing voice to the study and practice of Tai Chi."

—**Gene Ching, associate publisher of *Kung Fu Tai Chi* magazine & KungFuMagazine.com**

"Arthur Rosenfeld has written the most accessible book on Tai Chi I've seen. Its benefits are scientifically proven, and I'll be recommending this to my patients young and old."

—**Mark Lachs, MD, MPH, professor of medicine, Weill Cornell Medical College**

"Rosenfeld's book will improve your health and your mind. Easy and fun to read, it is filled with uplifting stories, lots to make you think about the world and plenty of easy-to-follow practical fitness advice. A delight."

—**Graeme Maxton, bestselling author and fellow of the Club of Rome**

"After my own decades of attempting to convey in ordinary English the deep and subtle insights of the Taoist traditions, I can appreciate the masterful contribution Arthur Rosenfeld had made with his *Tai Chi: The Perfect Exercise*. He brings sharp clarity to a subject too often shrouded in mystery and confusion."

—**Guy Leekley, author of *Tao Te Ching: A New Version for All Seekers***

"Arthur Rosenfeld is one of the most special and genuine voices in the arts today. Not persuaded by fame, attention, or self-congratulatory actions, he walks a path that is unique, winding, and full of discoveries, surprises, and truth, not just for himself but for those lucky enough to align themselves with him."

—Del Weston, martial artist, producer, writer and director

"Arthur Rosenfeld is rightfully one of the foremost Tai Chi masters in this country if not the world. This mastery has spiraled into his writing. This book has illumined my Tai Chi practice. It also offered fresh teaching examples in the areas of breath and energy that I can share with my students. I'm highly appreciative of his contribution with this book."

—Mitchell Doshin Cantor, Sensei of The Southern
Palm Zen Group

"Arthur Rosenfeld's new book, *Tai Chi: The Perfect Exercise*, breathes new life into the old saying *bun bu ichi* (the ways of the sword and those of the pen are one). It's extremely rare to find a martial artist whose practical expertise and martial insight are paired with literary elegance, enthusiasm, and rich experience.... Lucidly organized, elegantly written and filled with the types of insights that are only too rare in this genre.... The author's mastery of clear and accomplished prose...make this a valuable and mature meditation on the virtually limitless depths of this art."

—John Donahue, bestselling author of *Enzan*

"Rosenfeld's *Tai Chi* is as unique a contribution to the martial art as Bruce Lee's *Tao of Jeet Kune Do* was to his. This muscular work weaves history and modernity with philosophy and combat to create a tapestry that transcends all disciplines. *Tai Chi* will travel with you regardless of where you go and regardless of whether you take it."

—Cameron Conaway, author of *Caged: Memoirs of a Cage-
Fighting Poet*

TURTLE
PLANET

TURTLE PLANET

Compassion, Conservation,
and the Fate of the
Natural World

YUN ROU

Coral Gables

Published by Mango Publishing Group, a division of Mango Media Inc.

Cover Design: © Janelle Rosenfeld
Layout & Interior Design: © Morgane Leoni

For permission requests, please contact the publisher at:
Mango Publishing Group
2850 S Douglas Road, 2nd Floor
Coral Gables, FL 33134 USA
info@mango.bz

For special orders, quantity sales, course adoptions and corporate sales, please email the publisher at sales@mango.bz. For trade and wholesale sales, please contact Ingram Publisher Services at customer.service@ingramcontent.com or +1.800.509.4887.

Turtle Planet: Compassion, Conservation, and the Fate of the Natural World

Library of Congress Cataloging-in-Publication number: pending
ISBN: (print) 978-1-64250-271-8, (ebook) 978-1-64250-272-5
BISAC category code FICO25000FICTION / Psychological

Printed in the United States of America

For Professor Richard Malenky—researcher, environmentalist, and, perhaps most importantly, teacher extraordinaire. Thank you, old friend, for pointing me straight at what matters, all those many years ago.

And for Janelle, queen of all spirit realms, especially my own.

"The greatness of a nation and its moral progress can be judged by the way its animals are treated. I hold that the more helpless a creature, the more entitled it is to protection by man from the cruelty of humankind."

—Mahatma Gandhi

"Great Dao inhales and exhales. Monitor and guard sexual essence, breath, physical energy, emotion, money, and spiritual capital. Be sure there is more inhaling than exhaling, lest you flag, tire, sadden, go broke, forget who you are, and die."

—Immortal YIN, chelonian consort to the Great Sage, Laozi

"Behold the turtle, who makes progress only when he sticks his neck out."

—Bryant Conant

TABLE OF CONTENTS

———

WHY I WROTE THIS BOOK

Decades before I became a Daoist monk, I was born a seeker, always feeling as if I was staring at the surface of the pond, never willing to commit to a conventional life path for fear of missing out on what is really important. I've always sensed that we are creatures of time more than of space, and when it comes to time, we waste far too much of it. I believe we should question authority and doubt whether what we are told is true. I am wary of narratives that serve states or corporations and suspect the agendas that create those narratives.

One of those narratives that I find most disturbing tells us that animals are dumb and insensate, don't feel pleasure or pain, and that they, and the rest of Planet Earth, are here to serve us, sacrifice for us, and do our bidding. Decades of direct experience—floating eyeball to eyeball with a one-hundred-foot blue whale, dancing with the exquisitely deadly western Australian taipan snake, cuddling a hairless dog, teaching an African Grey parrot to talk, feeding a piranha, motorcycling with a California condor flying not far off the top of my helmet—tells me that this is a pernicious lie. In fact, I very much believe, as aboriginal people have for millennia, that there is a whole universe of *animal* experience and consciousness that stands separate and apart from *human* experience and consciousness. Western science is waking up to this reality, too. Even as we continue to torture them, butcher them, and drive them to extinction, more studies show that even animals with brains quite different from our own demonstrate consciousness, feel emotions, and possess intelligence.

Included in the standard narrative that denies this fact is a hierarchy of life, from low to high, bad to good, with human beings at the top. Given how much better other animals behave than we do—we corner the market on torture, trafficking, genocide, and other equally charming behaviors—this hierarchy is as ironically twisted as a molecule of DNA. It is also a relatively recent conceit. Rock paintings, oral traditions, and archeological evidence tell us that Paleolithic hunter-gatherers and Neolithic pastoralists respected other living creatures more than modern humans do, giving them their ceremonial and religious due even when eating them or harnessing them as organic machines. It is only as we violent monkeys have expanded in population and slaughtered more and more of our fellow creatures to satisfy our appetites that we have formalized the distinctions between humans and other animals, even drawing upon the science of evolution to justify the maltreatment of other species of sentient beings, and, of course, different races of humans as well.

If we were honest about this hierarchy—which like our gods exists only between our ears—we would upend it and place ourselves at the bottom. Had we done so in centuries past, Genghis Khan's generals would not have been able to train their foot soldiers to butcher innocent men, women, and children by convincing them that their victims were subhuman, and Adolf Hitler's commanders would not have portrayed Jews as rats so as to urge their underlings to acts of shocking and irredeemable cruelty. Again, it is this false hierarchy that has allowed us all to silence our conscience in the face of factory farming and to dispense with our moral compass in the face of the wholesale rape of one beautiful, natural community after another.

Better than switching around the rungs of the ladder, I say let's dispense with the ladder altogether. Hierarchies require judgments and judgments require distinctions, all things that Daoism, the religion in which I am a monk, regards with trepidation. Indeed, the major works of the Daoist canon all proscribe categorizations of all kinds, teaching that the minute we begin separating things one from the other, we separate them from ourselves, and that such a separation is the precise cause of the existential angst that so plagues modern culture. Only by constantly reminding ourselves that we are part of a coherent and indivisible whole, a continuum that runs from less than quarks to beyond galaxies, can we see the universe clearly and regain our sense of wonder.

The idea that we should do the right thing because doing so will benefit us has always offended me. I think it constitutes settling for a low moral level and that it sells people short. I don't believe that rampant narcissism is our fate, and I bridle at the idea that the only way to drive people to a higher moral position is to frame it as enlightened self-interest. I believe that deep down we all care about each other and the world, even though it may not always appear that way. Accordingly, I am loath to add "what's-in-it-for-me" material to this book. Nonetheless, the planet is careening toward total ecological collapse. If we wish to preserve the natural world as we know it, we must discard the idea that we are any better or worse than other animals, indeed that we are in any meaningful way different from them. We must stop deriding the term anthropomorphic (attributing familiar, human qualities and characteristics to animals) and face the fact that like us, animals feel pain and pleasure, bond to each other, hold the concepts of self and kin, and possess a will to live.

One way to effect this change is to become more sensitive and aware to the behavior of animals. Another way is to give them voice. In this book, I have done the latter for a familiar group of animals not only sorely suffering from human adventures on Earth, but also long regarded as voiceless. That group of animals is turtles. In choosing them for this literary adventure, I am hoping both to shine a light on the full spectrum of life and our own place in it, and to stimulate a global compassionate awakening.

WHY I CHOSE TURTLES TO TEACH ME

There is something quintessentially earthy about turtles. Perhaps it is that they are low and slow, although some can move quicker than we can; perhaps it is because they are generally silent, though some are quite vocal in love; perhaps it is because they are at once enduring and helpless, strong and weak, flighty and fierce, exploited but unknown. Perhaps it is that individually they live longer than we do and are therefore capable of perceiving the foolish foibles of each of our lives, though maybe it is more because, as a group, they arose before our own tree-shrew forebears, bore witness to the rise and fall of dinosaurs, and thus see our species in a geologic context we will never comprehend.

Turtles sometimes embody wisdom in literature, cartoons, television, and film, a wisdom born of both longevity and suffering. I look at them with both admiration and compassion, the first for their dogged, determined persistence, the second for their plight. Most folks don't look to them at all. Rather, they unthinkingly destroy their habitats, eat them in soup, grind them into potions, drown them in fishing nets, and even purposely run them over on the road. It is the fact that most people will not even pause to dignify turtles with a glance that makes these denizens of Earth's dark and unknown spaces such a perfect symbol of our dubious relationship with nature.

Turtles entered my life when I was nine years old and never left. I saw early on how a turtle in a pond or stream or river or sea could break water, take in what is going on above the surface, and then dive back down to a secret, but fundamental, world that human beings would never know. I envied them that ability—which for humans requires discipline, devotion, effort, and qualified guidance through esoteric waters, but for turtles comes naturally and with neither stress nor strain. I became fascinated with them. They connected me to nature at a time when I lived in an apartment building in a concrete jungle whose only trees were planted in ordered rows and whose clouds were mostly punctured by the radio antennae on top of skyscrapers.

Some of the species of turtles I knew as a child growing up in a Manhattan apartment are now functionally extinct, and even those that persist are hardly common in the wild. At the time of this writing, turtles as a group are the most critically endangered of all vertebrates. Early on, I had no idea of the threat they were under and no sense of contributing to their demise by participating in a pet trade that mortally drained wild populations. I simply wanted to be with them, to feed them, change the dirty water in their tanks, and see them blink in evident pleasure at their renewed world, fresh, clean, and clear. I wanted to watch them chase crickets, devour earthworms, suck down fish by making a vacuum cleaner of their throats, scoot joyfully through the water, stick periscope-like noses up for a breath of air, then retreat again to hide under aquarium gravel. In tending to them, I partook in a respectful, natural exchange of energy with the panoply of nature's other sentient beings. Now, after more than fifty years of working with individual turtles representing a third of Earth's extant turtle species, I am stunned by how many have been lost and how precarious is the position of the few that remain.

In choosing to write about them in this book, I aim to help stem the tide against them and to draw upon an ancient tradition to help reframe the environmental, social, spiritual, and cultural problems turtles face. That tradition is Daoism, the religion in which I am a monk and which is perhaps the first coherent form of environmentalism. Turtles and Daoism became intertwined in Neolithic times in the part of the world later called China. In those early days, shamans—interlocutors between our world and the next—burned the bellies of river turtles with red-hot pokers until they cracked, then read the resulting lines the way back-alley fortune tellers now read cards or tea leaves.

In one legend of the late Neolithic period, right before the dawn of Chinese dynasties, a turtle arose from beneath the surface of the Luo river right in front of Fuxi, an analogue to the biblical Adam, revealing a combination of broken and unbroken lines that would later underpin the *Yijing* (I-Ching), the divinatory urtext of Chinese philosophical culture. In another similar tale, the legendary Emperor Yu, who is reputed to have ruled wisely and justly from c. 2123–2025 BCE, was supervising the building of a dam on the Yellow River when a giant turtle surfaced before him. This turtle, called Hi, bore a message on its back:

4	9	2
3	5	7
8	1	6

When Yu took a closer look, he realized that the numbers in the square had a specific property, namely that in every direction, they added up to fifteen. This diagram is important in *feng shui*, an art used all over Asia to design and build homes, offices, buildings, gardens, and even cities so as to best potentiate stability, prosperity, health, fecundity, wealth, longevity, and material riches. Today, stone turtles, often with stone snakes on their backs, greet visitors in Daoist temples all over Asia, and real live turtles swim in Daoist temple ponds.

Turtles are reptiles. That means they are related to snakes and lizards and, more distantly, to crocodilians, but they are far from slimy-skinned amphibians like salamanders and frogs. They reproduce by copulation and internal fertilization, lay eggs, and can have quite elaborate courtship rituals. Most are love-'em and leave-'em types, but some stay around to protect their offspring. Like other reptiles, they have lungs, not gills, and possess a bony spinal column and senses that are familiar to us, along with a couple that aren't. Among the latter are the ability to detect electrical activity in water, to pick up vibrations in the ground using their shells, and perhaps the ability to sense specific airborne chemicals, especially pheromones.

Turtles are also poikilotherms, meaning that they depend upon external sources for the heat needed by their metabolic processes. They can be cold to touch when that heat is not forthcoming, leading to the misnomer "cold-blooded." Actually, their blood can be warmer than ours—whilst basking in the desert sun, for example—and in certain situations, some can even generate a little bit of their own body heat. Contrary to folklore, turtles are not particularly slow. Because of the limitations of their physiology, most are not able to sustain fast movement for long periods of time but can get up and move quickly in a short dash or swim. Giant sea turtles can even traverse entire oceans at a clip that only a high-speed submarine could match.

The turtle's defining characteristic is its shell. In most turtles—we will meet some exceptions in the fables that follow—the shell's surface layer is comprised of keratin, the same stuff as human fingernails and hair, and as the horn of a rhinoceros, too. This keratin is divided into sections called scutes, which are basically giant versions of the scales characteristic of all reptiles. The upper shell, the carapace, is joined on each side of the turtle to the lower shell, the plastron, by a bridge.

Beneath the scutes lie bony plates that approximate but do not exactly match the scutes above them. One unique feature of the turtle skeleton is that the shoulder girdle is inside the rib cage. If we were built that way, we'd be able to bring our ribs up to our ears when we shrugged.

Turtle shells range in shape from flat and streamlined to domed and unwieldy, the better to discourage predators such as alligators and wild dogs from crushing them for a meal. Some turtle shells close up so completely that only the most powerful and persistent predator can get inside to the soft parts; others have shells that are quite reduced to facilitate climbing or crawling about in the mud. Turtles with less protective shells typically make up for their deficits with strong jaws, a sharp beak, and a nasty attitude.

Speaking of attitudes, most turtles get along, at least with each other, and can live in large communities of different species. They have some surprising abilities, too. A few can spend an entire winter under the frozen surface of a lake or pond, reducing their heartbeats to nearly undetectable levels, and extracting what little oxygen they need through skin in their throat folds or urogenital area. The North American red-eared slider, *Trachemys scripta elegans,* has the most complex color vision system of any vertebrate, with seven different types of color-sensitive cone cells in the eye. Some turtles, while not nearly so eloquent and voluble as the ones we will encounter in the tales that follow, have a significant vocabulary. The Australian northern snake-necked turtle, *Chelodina oblonga*, uses growls, grunts, chirps, and more to communicate to other members of their species in murky, underwater conditions. Other tortoises vocalize loudly when mating.

Turtles inhabit Earth's most intimate spaces. So alien to us and yet so familiar that we routinely take them for granted, turtles remind us that despite our preoccupation with finding intelligent life out in the cosmos, such life is right here beside and below us. They remind us that we are most definitely not alone, though that is something that could also be said of many, many other types of animals. The real reason I chose them is that of all the creatures that crawl, fly, swim, run, slither, burrow, and climb in and through our world, they are the ones closest to my heart and the ones I know best.

DAOISM, MEDITATION, AND HOW A MONK TALKS TO TURTLES

Popular with Chinese elites since the fifth century BCE, Daoism can be seen as a religion, a philosophy, or both. Its fundamental permeability makes it nearly inextricable from the Chinese culture that birthed it, although right from the start, it proposed looking at the world in a way quantum physics, game theory, systems theory, and medicine all came to echo millennia later. Perhaps the first coherent environmentalism and argument for sustainable living, Daoism also foreshadowed Deep Ecology, a scientifically dubious but emotionally appealing philosophy first advanced in 1973 by the Norwegian Arne Næss and later popularized by British author James Lovelock.

The word Dao suggests a path or way, yet it is neither an entity nor a verb. Perhaps it is a process or presence, though Laozi, greatest of all Daoist sages, warns us not to confuse the word Dao with Dao itself. In the same sense that the word moon is not the heavenly orb, Laozi declares, "The Dao that can be spoken is not the eternal [real] Dao." In fact, it is easier to say what Dao is not than what Dao is. It is neither goodness nor God, though some see it as benevolent. Perhaps it is easiest just to think of Dao as nature, itself a complex and multilayered concept. Venerating nature and its workings, Daoists see all living things as part of the same vast, ever-changing tapestry of matter and energy. We believe that in our original state, unsullied by propaganda, commerce, and the bewildering array of agendas forced upon us in our modern lives, we all have access to the direct, spiritual experience of "something going on."

Westerners know more of Daoism than they realize, as George Lucas clearly drew upon Daoism when creating his wildly popular Star Wars franchise—so much so that late entries into that filmdom even feature overtly Daoist symbols. In that fictional world, the rebels represent Daoists while the empire represents the competing, more rigid philosophy of Confucius, which relies upon strict codes of moral conduct, government regulations, and inviolable social roles and positions. On the surface, Star Wars chronicles the struggle between Jedi masters who understand universal forces and wield swords like kung fu heroes on the one hand, and generals and soldiers who embrace guns and bombs as the tools of tyranny on the other. On a deeper, more real-world level, it is the conflict between lovers of nature and lovers of law and order, between free-thinkers and conformists, between those who embrace evolution and change and those who fear and resist both, between, dare I say it, literal (not political) liberals and conservatives.

What Star Wars fans do not know is that in addition to conjuring a love of nature and an Eastern view of the world, Daoism offers physical practices, rituals, techniques, poetry, literature, music, and arts, all of which brim with exoticism, wildness, insubordination, revolutionary evolutionism, a deep love of truth, individuality, diversity, and a nonjudgmental, egalitarian world view. Today's flesh-and-blood Daoists may not carry light sabers, but they have little tolerance for submitting to authority nor to conforming to societal norms. Rather, they live balanced, spiritually rich lives, embracing compassion, frugality, and humility, Daoism's three spiritual treasures. The ideal Daoist life is one of free and easy wandering, whilst being maximally effective with minimum effort. A term for this attitude in action, or nonaction, is *wu wei*.

Change is integral to Daoism and is represented by the yin/yang symbol seen on everything from surf-board-toting pickup truck windows to hip clothing. Yin and yang are terms commonly used to describe forces, states, energies, or qualities and are typically portrayed as opposites. More accurately, they exist only inside the circle that encompasses them and are, together with that circle, a symbol connoting motion and constant change. The symbol, properly called the *taijitu* is best seen as a movie rather than a still image. In that movie, yin and yang exist only in relation one to the other and are in a constant state of exchange, one perpetually replacing the other. Think of the way day replaces night, the two changing places at dawn, and how male and female each contain a bit of the other. This rich tapestry of constant change is quintessentially Daoist and is seen as an accurate portrayal of the way nature "works."

Although many schools and lineages people the history of Daoism, and many great teachers were themselves students of even greater ones, it is also true that many highly respected masters received their education in a fashion recognizable to followers of other faiths. As the God of the Old Testament is said to have transmitted the Ten Commandments to Moses or the Archangel Gabriel planted in Muhammad the seeds of the Quran, ascended masters in the Daoist tradition, essentially gods, sometimes contacted individuals they deemed worthy of profound lessons. Such deities include Lord Lao himself (a deified version of the author of the *Daodejing*) and the so-called Eight Immortals, some of whom were purely fictional characters while others were mythical versions of actual historical personages.

We cannot share the exact experience of receiving such transmissions, for we cannot even get inside the heads of people right next to us, never mind people long dead. What we do know, though, is that colorful adventures, verbal instructions, visions, and more were grist for the mystical mill. Such mystical experiences have been linked to extended and rigorous meditation sessions, sometimes in groups in a sylvan setting, sometimes alone in a cave in the dark for weeks or even months on end. Although few Daoist teachers these days speak much about it, some early Daoist shamans may also have used consciousness-altering drugs, either marijuana or frank hallucinogens, as part of their consciousness-raising rituals.

Though present in the mystical traditions of other religions as well, exploring the types and nature of consciousness is a quintessentially Daoist pursuit. Indeed, Zhuangzi, one of the most famous and most beloved of all Daoist sages—and China's first novelist, too—wonders aloud, in a famous passage in his eponymous work, whether he is a man dreaming he is a butterfly or a butterfly dreaming he is a man. The fact that he chose a butterfly for this inquiry, as opposed to a prince or warrior or king, in turn reveals just how important the natural world is to Daoists, and just how strong the continuous link is between all living things.

The information transmitted during these trances sometimes pertained to the circulation of energy in the body, sometimes to the nature of reality, and often to achieving immortality, still a major goal of religious Daoist practice. Once a religious figure received this information, he or she was likely to share the new information with their flock. This sharing involved sermons and so-called spirit-writing (*fuji* or *fuluan* in Mandarin Chinese). Spirit-writing is deeply ingrained in Daoist tradition and is a major source of classical Daoist wisdom to this day. In some writings, the author holds a question or goal in mind and seeks an answer or guidance. In other writings, the author empties his or her mind and waits for the deity's input. Both methods are legitimate, and each has their potential pleasures and pitfalls, benefits and limitations.

In keeping with the spirit-writing tradition, which has greatly inspired me over the decades of my Daoist career, I offer this book. Rather than receiving transmissions from Lord Lao or any of the classical Eight Immortals, I am taught by immortal turtles. Sometimes I ask them questions and sometimes they tell me what's on their minds. When the first of these chelonian masters appeared to me, I could only feel a great sense of "rightness" to the experience, for shelled reptiles had been with me all my life. I felt, and still feel, no more deserving of these marvelous transmissions than, I suppose, any historic Daoist sage did when confronted by a luminous teacher offering blessings. All the same, I drank the wisdom like a fine wine and am hopeful—in part because of the content and in part because of the context—that sharing this material will benefit the world.

Did I choose turtles or did they choose me? Did I self-generate these conversations, or was I literally visited by Daoist immortals in turtle guise? I leave that up to the reader to decide. I don't suppose the literal truth of such journeys or visitations is of any more importance than the historicity of our ancient religious leaders, even though some fundamentalists get all excited about that question. To me it is, and has always been, the message, the *lesson*, that counts.

WIRED FOR STORY

I stand in meditation as I always do, in a cosmopolitan park, in the shade, near a lake, motionless as a tree, my feet shoulder-width apart, my eyes closed, my hands folded over my navel. Usually in these sessions, which I have been doing for decades, I exist in both interior and exterior worlds, my thoughts bouncing between the two. The frequency of that bouncing is not particularly related to the success of my meditation, for the idea of forcing my mind to do anything is antithetical to my tradition. I might, therefore, feel the ache in my thigh muscles from standing so long while hearing the zing of bicycle tires on the path nearby. I might feel a chill on the back of my neck because the wind has picked up while at the same time hearing the multilingual chatter of nearby park-goers. I might get a whiff of beer and notice a heavy, gauzy feeling in my palms. I might feel a drop in barometric pressure as a storm approaches while noting that my respiration has dropped to a mere two breaths per minute.

This time, I experience something new. It is a form of noticing, that's true, but it's noticing an absence rather than a presence. Rather than something changing, shifting, or appearing, I notice that I suddenly don't hear anything at all. It's as if I've got water in my ears, but I haven't been swimming. It's as if the familiar buzz of an air-conditioner or refrigerator has suddenly abated, as if during a power failure, say, leaving me with a sudden awareness of silence. It's as if I've entered a sensory deprivation tank, but I have not; I'm still here, standing in the park. The silence is disorienting. Heavy. Unnerving.

It makes me realize just how much I depend upon binaural hearing to locate my place in the world, to substantiate my presence in space. Honestly, it's a bit nauseating.

To reassure myself, I open my eyes. My intention is to reground myself, to remind myself where I am, and then to close them again and resume my meditation. Instead, when I loose my lids, I find myself in a field of frosted light. There is no park. There is no lake. There are no tourists, pastoralists, or revelers. All is quiet and glowing. I blink and try again. Still nothing. I wipe my eyes. I can feel the pressure of my fingers on my lids and see the darkening my proximal flesh causes, but as soon as I stop rubbing and open my eyes again, I'm back in the silent, frosted world.

I feel a bit of panic coming on. I wonder if I'm having a stroke. I inhale deeply through my nose, hoping to pick up a whiff of something familiar or at least orienting. That's when I realize that my sense of smell, too, has fled, or at least is offering no hint that I am any longer in my familiar park. It's not that I smell nothing, but rather that I smell something vaguely familiar yet most definitely not the smell of my meditation park. Sighing, breathing, trying to steady myself, I focus on the pressure of the ground underfoot. There's always feedback to be found there, even if it's my toes being squeezed too tightly by my shoes. Usually, in this favorite spot of mine, a couple of tree roots make themselves known to me even though they are well underground. How do I usually know they're there? Maybe it's some energetic effusion. Maybe it's that my weight compacts the overlying soil just enough for my sensitive monk's feet to feel them. In any case, this time they are not there. In fact, the ground underneath feels slippery, hard, unyielding.

In fact, it feels wet. I open my eyes again, look down, and see that I'm standing in water up to the level of my calves. My monk's slippers, along with my white leggings and the bottom hem of black robes, are wet. Not knowing what else to do, I start to walk. That in itself is a first. I've meditated in the rain before, accepted being a bit sodden rather than break my mental momentum, but actually moving while in a meditative trance has only happened when I was doing tai chi, and in that case, I was very much in the waking world. After a few steps, I discover I'm headed up some kind of rise. I proceed using a martial arts technique called inch-stepping, in which I lift the front foot and slide forward, propelled by the rear. Keeping my weight on my back foot this way, I minimize the risk of slipping on the smooth surface below me.

This seems to go on for a while, and as it does, I don't think about the spiritual transmissions of Daoist immortals to deserving sages. Instead, I just wonder where I am and wonder again, at least fleetingly, if I'm perhaps suffering a stroke. I've heard that pathological cerebral events can cause this kind of synesthesia, this kind of strange experience. Maybe, it occurs to me, no time has passed at all and nothing has actually happened other than the strangling of the blood supply to some tiny portion of my brain.

It's at the end of those thoughts that I suddenly realize where I am. I'm not in the park, but neither am I in Stroke Land. I am, in fact, standing inside one of those little plastic turtle bowls that used to be sold at the circus (along with baby turtles) or featured in the back of *Mad Magazine, National Lampoon,* or comic books I used to read, right alongside brine shrimp billed as "Sea Monkeys." The difference is, obviously, that it's a bowl big enough for me to stand in, and it features, I now see, a monk-sized plastic palm tree. I experience a small frisson of excitement. This, truly, is a new level of meditative experience for me.

It is when I relax into my surroundings that I see a turtle snoozing beneath the plastic palm. She raises a white eyelid and regards me.

"Finally and at last," she says in a twee voice.

I recognize her to be a red-eared slider, the most common and widespread of all the world's turtles, and precisely the one most people put into these little plastic houses before eventually flushing them, dead or dying, down the toilet. Her carapace is green. Her plastron, reflected in the plastic below us, is bright yellow with dark figures. They might be intertwining Renaissance-painting nudes.

"What is this? What's going on?"

"You're a Daoist seeker, are you not? A monk?"

"I am."

"And you have questions that burn in you and a desire for both deeper and broader understanding of how things are?"

"I do."

"And you're familiar with spirit-writing? With the phenomenon of receiving transmissions, via trances or travels, from enlightened immortals, and then sharing the information?"

I am momentarily thunderstruck. "You're saying—"

"Yes, yes," the slider says impatiently. "It's happening to you."

"But I thought—"

"What, that you'd rise up on a cloud to a heavenly garden and eat peaches with bent-over old men? There are many immortals besides the proverbial eight, you know, and we appear in different forms according to what is expected, according to what will get the job done."

I feel a tremendous joy arise in me. "Is this really true?"

"As true a transmission as any," the slider answers. "The first, but not the last for you."

"How many will there be?"

"Are you sure you want to know that? Wouldn't you rather live in the moment, never knowing when another will happen, preserving a state of joyous expectation for the rest of your life?"

"Never mind how many," I say.

"What I will tell you is that we immortals will all come to you in the forms of turtles. At least for now."

I find a dry spot on the hard, clear plastic under the tree and sit down next to her. "Here?" I ask. "In this plastic bowl? I'll receive all my transmissions here?"

"That would be boring. Besides, this sterile, plastic world is a terrible place."

I shift positions, trying to get comfortable but unable to manage it. At every angle, something unyielding seems to find a soft spot in me, a buttock, a hip, an ankle, a knee. "It is," I say. "I hate plastic. When I wear plastic shoes, my feet sweat. Drinking from plastic bottles makes me feel vaguely off. And it's so hard to get comfortable here."

"Don't I know it. Turtles have their hard parts, but they have their soft parts, too. Imagine the millions of us who suffered such a cruel fate over the decades they were sold along with these terrible plastic prisons."

"You're saying all turtles are immortals?"

"Don't be dull. Among the famous human Eight Immortals (we immortals inhabiting turtle bodies think those eight are overrated, by the way) there was one who was transgender, another who was a poet, another a warrior, another a cripple. Does that mean all such people are immortals? We appear as we must to those who need us. Anyway, I'm the one we turtle immortals agreed would be the first to meet you and explain that our transmissions to you are going to be in story form."

"Story form? I haven't heard of that before."

The slider does a classic turtle stretch, spreading her claws and extending her neck and limbs to show the beautiful patterns of her skin including the oblong, tell-tale red patches behind her ears.

"If you're referring to tradition, I remind you that story is what distinguishes human beings from so many other creatures and has been an essential part of your biology since your very early days. Long before there was what you call Daoism, your ancestors would sit around fires and share tales of nature and of doom, of excitement and sex and kindship and war. Those tales established what you call archetypes, the building blocks of your culture, those ideas and values and principles that you all handed down from one generation to another, mostly inside clans, even before there was any kind of writing. In those archetypes were to be found the rules that made your societies work, along with the paragons and saints, sinners, too, who served as models for what to do and not to do, how to live and how not to live. Stories in those days were consummately relevant to the listeners. They addressed everyone's hopes and fears and desires, their longing to believe they went on after dying. Storytelling was the chief means of bonding when bonding meant survival, when the only other thing worth doing was to make babies or hunt food or sit quietly and observe the unfolding of nature."

"So is that what we're doing now? Bonding?"

"We're preparing you to accept the transmissions to follow. The lessons. You'll soon discover that each session will not only have a different message and content but will unfold in a unique and interesting way for you. Don't get me wrong. They may be challenging, but that is the nature of spirit work."

"You're taking about spirit-writing."

"Yes. And I'm only here to remind you of the power of the narrative you will derive from your experiences and share. Narratives define reality for human beings. If you lose your narrative, you lose everything, and the ultimate source of narrative—the wellspring that is always there for you no matter how lost you think you are—is nature. Despite widespread fantasies about a cloud-sitting beard-stroker running the show, nature is actually all there is, all there ever has been, and all there ever will be."

"Very Daoist," I say.

"Yes, indeed."

"Would you tell me where we are?" I ask. "This plastic place? You must have chosen it for some reason."

"You really don't know?"

I frown. I don't know quite what she means.

The slider sees my confusion. "You had it for a moment," she says. "You recognized the smell."

Once again, as I did a few minutes ago, I inhale deeply. There is something there. Something familiar. Not one odor but a complex. I search my memory. My grandmother's potato knishes? Carpet soap? Furniture oil? A shaving cream I used at puberty, something special that my father gave me that numbed my tender young skin against the razor? Just as I'm beginning to puzzle it all out, a shadow appears between me and the turtle, cast by whatever source of light illuminates this place. The shadow moves, resolving first into the head of a wolf, then a dancing chicken, then the jaws of a crocodile spread across the sky above me, then the tall ears of a prancing rabbit, and finally, double-sized, a flying, predatory dinosaur with a probing, fearsome head.

"Someone's making hand animals," I say.

The Red-ear's beak does its best imitation of a smile. "And who do you suppose it is?"

I see the boy right before I answer. He's charmingly buck-toothed, with bright, kind eyes and long eyelashes, a generous nose, thin lips, and a thatch of thick, black, much-missed hair. He's humming as he makes the hand animals, and, before long, he's crooning a Bob Dylan tune. *I've been out in front of a dozen dead oceans.* His voice isn't bad. In fact, it's rather good. On pitch. Strong. I take some pride in it, because he is *me*—perhaps half a century ago.

"Remember anything?" the Red-ear asks.

"Of course. I've always loved that song."

When the boy bends closer, the world behind him resolves into my childhood bedroom, the first place where I kept and loved turtles. The white bookshelves are there, full of books about animals, and there are photos, my own, of salamanders, frogs, snakes, and landscapes that calm me. Since I'm obviously in my boyhood home, all those smells now make perfect sense.

"Do you remember that it was your allergy to animals with hair that first drew you to turtles as pets?"

"Of course I remember. I've got a Mexican hairless dog now."

"You killed a lot of your turtles," says the Red-ear. "Some died of lack of calcium, their shells paper thin. You flushed them down the toilet. Others died because you kept them too cold, and they never ate quite enough. One died because you thought he had escaped, didn't realize he was buried in the dirt, put the tank away, and let him waste to nothing in your clothes closet. You killed some more when you used too much poison, or the wrong kind, when trying to rid them of parasites."

"Stop," I say.

"A few burned in that accident where you left scalding hot water running while they teetered in Tupperware on the edge of the sink. That was the worst. They suffered so much. And you. You cried and cried."

I find myself crying now, even as the young version of me continues to sing. Tears for the animals nobody cares about, tears for all the turtles dead on the road, tears for the ones that ended up in my own toilet, tears for my own ignorance of what my pets needed, tears for not understanding their needs, feelings, and worlds.

"Please," I say.

"None of it was intentional," says the turtle. "We know that or we wouldn't be here. We know how much you love our kind."

I let out an anguished howl. "I'm so sorry! I didn't know better! I've spent thousands of hours taking care of turtles since and I've done a much better job!"

"We know that, too. You've redeemed yourself a bit, and this spirit-writing will redeem you further."

Together, we watch the boy with all that beautiful hair dance around his bedroom, circling the little plastic lagoon in which we huddle. I think about how fleeting life is, and how we can do nothing with the days we have but take care of each other and live in daily, hourly, minute-by-minute grateful appreciation of nature.

"What do I need to do to make it happen?" I ask.

"Stand in meditation. Wait patiently. One of us will appear. Each visit will be different."

"What can I expect to happen?"

"Well, all we really know of all those famous episodes of spirit-writing is the result. Neither of us was there for the process. We don't actually *know* what went on between the immortal and the sage, between the deity and his or her recorder. We can *presume* to know, figure there was a vision or a voice or some other kind of visceral experience, but in truth we have no idea."

"That doesn't really help me prepare."

"Ok, how about this? In some visits, a turtle immortal will *tell* you things you need to know by speaking of them directly. In others, the teacher will *show* you the lesson. And in other cases, you will be learning or living right along with your teacher. How's that?"

"Not very specific," I say, half worried and half excited.

"But you agree?"

"Oh, I agree," I say. "In fact, I'm honored."

"As you should be," says the Red-ear. "Writing the things you will write is the Daoist path to immortality."

"That sounds great," I say, "but I actually want to help turtles."

Yet because my old bedroom has quickly gone fuzzy, and I am back in my meditation park, I'm not entirely certain she hears me.

BE LIKE WATER

Once again, I stand in Daoist meditation. As usual, my arms are out in front of me as if I'm hugging a tree, my elbows and shoulders are down and relaxed, and the tip of my tongue on the roof of my mouth. Children laugh nearby. Geese squabble. Runners curse what the geese have left behind. I wait for something to happen. I wait some more. Gradually, the park recedes, and I find myself paddling through frigid ocean water, dark blue and forbidding. I take a few strokes and feel the pressure of the water against my hands. My fingers prune. A leatherback sea turtle, mistress of the open ocean, appears beside me. A gargantuan turtle the size of a small car, she has a ridged, teardrop-shaped shell, bulbous at the front and tapering to a point at the back. The head and limbs bear beautiful white snowflake patterns.

"I'm so happy to see you, Monk. I thought you might arrive too late to hear my story. What a shame that would have been."

"I'm breathing underwater," I say.

"You're fine. Daoist magic, remember? I'm the one in trouble."

It takes me a moment to understand what she means. Then I see it. A piece of fishing line, anchored down into the abyss, is wrapped around one of her front flippers. It oozes blood.

"You're stuck," I say.

"I am."

"So you can't get up to the surface to breathe."

"I can't actually. I have approximately eleven minutes to live. Perhaps a little more if I gentle my mind and drift with the current. If you don't mind watching me pass, you may stay with me."

I wonder what kind of immortal drowns caught in fishing line. I ponder the sort of lesson I can expect from spending the next eleven minutes watching a sea turtle drown. The Red-ear, perhaps the chelonian analogue of Lü Dongbin, the immortal leader of that famous gang of eight, informed me that the lessons I would experience would be unique and challenging. Be that as it may, I can think of few things I would rather do than watch this magnificent creature drown.

"Is there no way for me to help?"

"You can try," she says.

I swim over to the offending fishing line. The water is dense and cold. I ask the turtle where we are.

"In the plastic deathtrap of the North Pacific Gyre," she says. "Halfway between California and Hawaii."

"I've spent a lot of time in Hawaii," I say, trying to get my fingers between the flipper and the line. "The volcanoes are getting angry at all that pollution and overpopulation. It's not as nice a place to be as it once was."

"The islanders call the volcanoes by their goddess name, *Pele*," says the turtle. "Of course, she's seeking her revenge, though we Daoists would call it returning to *wuji*, to balance. Equilibrium. Humans have run amok."

"It's not just the eruptions," I say after clearing a gulp of salty water. I find I can't break the line with my fingers or gnaw through it with my teeth. "The island spirit of *aloha* is hard to find among the transplants, and there's a lot of racism and intolerance and traffic. Violence, too."

The turtle spins around in the water so she can see what I'm doing. "You're not going to get through it," she says. "If it's too strong for me and shark and tuna, it's too strong for your little fingers. The only way is spiraling. It's what galaxies do. It's nature's way of dealing with conflict."

I follow her instructions and by swimming around her trapped fin, begin to loosen the line.

"Speaking of conflict," the turtle continues, "Violence is one of nature's favorite ways to cull humans. Everyone knows she uses volcanic eruptions, earthquakes, floods, typhoons, mudslides, droughts, dust storms, and tornadoes, but many people don't recognize that all those terrible diseases are her work as well, as are religions that make you fight and kill each other, gender variations that reduce your reproductive rate, and, of course, the automobile, which takes so many human lives. She is the mistress of hate as well as the mother of great beauty, this ruler of ours, and mistress, too, of what you call war. Believe me, Monk, nature is ruthless in her campaign to survive you."

"You see all this from beneath the waves?" I ask, a bit incredulous.

"Oh yes," she says as I continue to unwind the line. "I see a great deal from the open water and from beachheads I have climbed, but of course I also hear from other turtles. Turtle Immortals band together, you know, and we've been watching humans for a long time. My cousins speak to me of your doings from their homes across the globe. We watch you from forests and jungles and rivers and streams, from deserts and mountaintops, too."

"I never knew," I say.

Her beak will not curve to her mood, but in the faint crinkling of the skin around her eyes, I detect an indulgent smile. I continue unwinding the countless layers of line that trap her, but she's made the problem worse with her own spiraling, and I worry she will run out of air before I can get free her.

"I've paddled to the Atlantic from the Gulf of Saint Lawrence to the North Sea, from the Cape of Good Hope to Labrador, she says dreamily. "I've greeted young of my kind off the beaches of Suriname and Guyana, Antigua, Barbuda, Tobago, and Gandoca and Parismina in Costa Rica. I've crossed to the Pacific and rested on the sands of Papua New Guinea and Gabon. Movie stars have watched me in California. I've been churned in the wake of a freighter off Malaysia, where once thousands of my kind gathered at the beach of Rantau Abang before the locals dug our eggs for soup, and kayakers have brushed past me in British Columbia. Once, years ago, I visited kin off the Nicobar Islands, but I never went back because I saw a ghost on a dune and beheld so many young of my kind fall to birds."

"Do your friends know you're stuck?" I ask, pushing away the cloud of tangled line I've thus far managed to remove.

"Of course. That's why they sent you."

As I kick my way around her with the line in my hand, I notice a graveyard of bones on the seafloor beneath me—remains of air-breathing creatures who could not reach the surface to save themselves—primarily the long spines, broad tails, cavernous ribs, bristles, and teeth of whales. Evidently cleaned by scavenger fishes and the scouring of the constant current, they glow white.

"How old are you?" I ask.

"I was 204 last year."

"You're so experienced. How could you get caught in this line?"

"The multitude of new challenges in this world tire me, Monk. The search for food in this raped ocean, the drift nets, the long lines, the hooks, satellite tracking, harpoons. All of my kind struggle to survive and meet and mate and breed and survive, but there are only so many times any one of us can escape life's traps and pitfalls."

"Weren't all those bones down there a clue?"

"I dove deep hunting jellyfish because I love the sting of their tentacles on my flesh and the satisfaction as they slide down my gullet. I got caught on the way back up. I indulged my desire as your kind does, and I paid the price as all of us do. In the end it is always our mistakes that kill us."

"Still," I say, as my work finally reveals her scaled limb, free of line, "you're so brilliant and this seems such an amateur's error."

"Brilliant," she scoffs. "Why do humans have such a romance with exceptionalism? The one thing you fear most is to be normal, to be regular, to be just another shell in the crowd. There is so much to celebrate in the very fact that life exists. Why focus on our differences rather than the marvel we share?"

"I see the similarities and differences of expressions of Dao," I say. "A flexible spine for me, a shell for you; a brain like a cat's in my belly, a sphincter that keeps out salt water in yours; a pelvic girdle that lets me stand and run, shoulders inside a shell for you; my brain for plotting books, your thousand-meter-dive lungs; smartphones for me, geo-positioning head magnets for you."

I finally free the line. All that's left is one of the hooks. I work it out slowly, seeing the pink and white of her flesh as I do, very much aware of how much what I'm doing must hurt her. "Now you don't have to worry about the killer whale," I say.

She turns to me, her giant beak not inches from my face. "What do you know of killer whales?" she asks, her voice unfriendly for the first time.

"I saw one chasing you right as you appeared to me," I say. "I saw his jaw open and I saw you flee in desperation into the nets. I saw him veer off before he, too, became entangled. I saw you choose a lingering death over a quick one."

She exhales a cloud of bubbles and heads slowly to the surface, her great flippers beating in tandem, the injured one not quite as strong, rendering her progress less than direct. "I refused to be food for him," she tells me. "I couldn't let him win after two centuries of eluding his kind in all their warm-blooded, arrogant cunning. They are like humans with tails, thinking they rule the underwater realm, seeking to supersede those of us who were here first, who have seen and understand things they never will."

"I'm sorry," I say, working hard to keep up with her ascent. "I meant no disrespect."

"I remember the first strokes I took after climbing out of my mother's nest on that beach in Sri Lanka. I remember crawling and stumbling and tumbling into the water and feeling the warm caress of that tropical ocean, back then free of the myriad chemicals humans have flushed into it. I remember what it felt like to fly through that water, and I remember every mile I've flown since. Being in the water, I try to be *like* the water. If your kind could do the same, if all of *you* could be like water, your lives would be better for it."

"Some of us have that ambition," I say. "Some of our great texts talk about it."

She waves her free forelimb dismissively. "Water gives life far more often than it takes it. If any of your kind actually *lived* the lessons of those texts—instead of solving mathematical problems and building torpedoes—you would thrive without destroying everything around

BE LIKE WATER

you. You would yield instead of standing fast. You would roll in and out like the tide, cyclically and without effort. You would understand that all things are the same, not different, and that the rights and wrongs to which you are so addicted are functions of perspective and dogma. All of you, each and every one, need to realize how we are all connected."

"People can be rigid," I say. "That much is for sure. Maybe my spirit-writing will help."

We break water. The wind is high, sending little whitecaps scudding everywhere. The mid-ocean swell is big, and I rise and fall as if in a skyscraper's elevator. She floats alongside me.

"Thank you for saving my life," she says. "That wasn't the best way to go, although in truth I am ready and will go soon."

I can't say exactly why, but when she tells me this, I start to weep. Maybe it's relief at feeling the sun again, strong on my face, maybe it's something else. She nudges against me with her shell, and I'm surprised to find that she's nearly as warm as the sun.

"Being like water means no attachment to outcome," she says. "Do you think each wave celebrates its own forming and then bemoans its return to the surface? Of course not. Instead of holding onto a great collective obsession with seeing your will done and preserving your life, you should follow water's example and avoid meeting force with force, understanding yourselves to be part of something larger, and accepting your end with equanimity, for there will surely be another beginning. It shouldn't be so hard. After all, water doesn't struggle."

"I'll pass that along," I say.

She raises her good flipper in what I suppose is a turtle version of a wave. I raise my hand to respond in kind before she sinks out of sight, but before I can, I find myself on terra firma once more.

I'm in the park and it's drizzling.

WE CHOOSE THE GODS THAT SUIT US BEST

There is a Latin culture festival going on in the park today, and people from Central and South America are parading, playing music, and singing songs in Spanish. They're having so much fun; it's terribly tempting to watch, but I'm there to meditate. I try using a nearby tree for interference, orienting myself so I'll be as unobtrusive as possible and so that the tree itself will cut the noise a bit. I finally manage to settle myself, to slow my heart rate and my breathing, and to prepare for what the Turtle Immortals may have in store for me today. While I'm waiting, I notice a foul and fetid odor. It's faint at first, but it grows stronger and stronger until I suddenly find myself underwater again, this time floating just off a muddy bottom.

"Nobody told me you were bald," says a stentorian voice.

I look for the source, but the water is turbid and topped with gritty foam, and all I can make out are waterlogged branches, old tires, and the rusted frame of a discarded car door.

"You have me at a disadvantage," I answer. "You can see me, but I can't see you."

"Flint River wasn't always this dirty. Hundred years ago, I could see the craters on a full moon just by crawlin' up on the bank for a spell. A year later, the crappin' it up began. That's Georgia for you."

The voice seems to be coming from below, but I still can't make out the source. I fan my hands to lower myself and take a few shuffling steps in the mud.

"Okay, but where are *you*, exactly?"

"River starts up by the airport in 'Lana. That's where the aviation gas gets in, and the oil. Chemical spills of the stuff used to take the ice off airplane windshields. Hey! You're about standin' on my goddamn head. If you weren't the monk I've been waiting for, I'd bite off all your toes. Look. Here's a fart."

A few bubbles appear. I look for a source but still don't see one.

" 'Fore today, I never did know a man's brains fall out with his hair," says the voice.

I'm tiring of the game when the turtle opens an eye. It's the size of a cup of coffee. I follow the line of his enormous head forward to his snout and then back to his neck and the borders of his shell. He's an alligator snapping turtle, an armored giant with a long, thick tail, the largest freshwater turtle in North America. Perhaps he weighs two-hundred pounds.

"I shave my head with a safety razor," I tell him.

"If you say so, Monk. You ever listen to Clifton Chenier?"

"You're asking me about a musician?" I say, confused. "The King of Zydeco?"

His enormous head comes out of its shell and stretches forward. "There you go! Now you're talkin' music. Know what I like best? Hearing that man play his blend of R&B and indigenous tunes on his accordion while I get a nice fat fish in my beak. Lure him in like this..."

He opens his mouth. In the gloom of the river, a red beacon beckons. It looks like a giant writhing bloodworm. I watch in fascination as a school of perch close in. One takes the bait and an instant later is lost in the maw. A spray of shiny scales floats away.

"I'm sure that's very effective," I say.

"These days, I'm lucky to find a fish with only one head," he says. "Since they've been dumping so much in the river, lotta the fish have two heads, three heads, even four. They've got too many tails sometimes, too, which means they don't swim right. That makes it easy for me to snatch 'em right up."

"You think all these mutations are from water pollution?"

"You know, I started out hardly bigger than your thumb. I've been feasting in this river since the beginning, eating ducks and armadillos, muskrats and beaver. I eat snakes, too, and smaller turtles. I even used to eat nutria because so many of them got out of those lousy traps the fur traders built."

"You eat turtles? Your own kind?"

"You telling me your kind doesn't eat each other? I love me a good map turtle on a cloudy day, a musky anytime. Delicious."

"People don't eat people."

"The hell they don't. I've seen them do it."

"You've seen cannibalism?"

"Tribes along this river. Not lately, I admit, but back in the day. They did terrible things, things I don't want to think about it at all. Spearing each other and painting their faces and bodies and cooking each other up in fires."

I sit down in the mud beside him. I try to convince myself I'm taking some kind of spa treatment. It's not easy to do. There's nothing pampering about this mucky bottom, and every time I move even a little bit, a cloud of particles rises into the water column. I don't know what the particles are, other than some tiny plastic beads, but I'm sure there's nothing that's going to help my health or clean my skin.

"Native Americans in Georgia were not cannibals," I say decisively.

"None of the others told me you'd be so sure of things you don't actually know. What *I* know for a fact is that monks are supposed to be humble. Now, you're here to receive transmission and not to argue, yes?"

I put my hands together in prayer and bow in his direction. "Yes," I say. "I apologize."

"Not only have I seen that, I've met the God of Music, too. I heard him. I met him. He was my friend."

I consider this for a moment. I know that a few of the legendary Eight Immortals of Daoism are musicians. I figure that since this turtle is an immortal himself, he must be referring to one of those. I ask him if he means the one named Han Xiangzi, because flutists consider him their patron and because he composed the famous piece, *Tian Hua Yin*.

"Bah," scoffs the turtle. "An amateur. And he's an immortal. I'm talking about a god."

"Lan Caihe then? A flutist also."

"That one sang ridiculous songs. A mere pretender. And again, one of the eight, not a god."

"He Xiangu then," I say hopefully. "She had a flute, too."

The snapper rises just a little bit onto his limbs. The act of separating his bulk from the riverbed creates eddies of mud. Glinting minnows spiral into the vacuum created by his absence. A large crawfish scuttles away. It has been hiding in the snapper's shadow too, for no predatory fish would dare venture close to those jaws, no heron or ibis dart would dare dart in. The snapper sees it and swallows it whole.

"Are you thick? I keep saying a god and you keep naming immortals. You think about those eight too much. Gods are to immortals as immortals are to regular creatures. More or less, anyway."

"Elvis, then?" I ask.

"Very funny. I'm not referring to any mere performer, though you could be forgiven for thinking there might be some holy water running through Elvis's veins."

I confess to being fresh out of guesses.

"Can't blame you for being ignorant," he says, slowly nodding his enormous head. "So many human beings are. You do know that there were cork trees along this river once. That wood was light as a feather. I used to climb up on broke branches and float along in peace, keeping an eye out for herons of course, and eagles. That's where I first noticed him casting a fishing line from a dock built out the backside of his father and mother's place. He was wearing a flat hat with a broad brim and dungarees and a church shirt, white with buttons. His feet were bare and there was a little key dangling from a chain on his neck. It would be a long time before I knew what that key was for."

"He appeared to you as a human?"

"We live our stories, Monk. All of us. I lived and died in this river. He came to me in the form I wanted him to, even though I didn't fully understand that at the time. That's what gods do. He used to roll dice on the dock, and I'd hear the clatter and come up and when he saw me, he'd take out his harmonica and play while he watched me. Later, when he got older, he found him a redhead girl with freckles—married her after a bit—and they'd sit on the dock with their toes in the water, which the girl said was brave on account of the fact that I so love to eat toes. She was right about that, but he knew I loved the music more, loved it so much I wouldn't risk hurting him or his kin. That girl played the violin like she was a robin and it was her egg. I'd come out for the two of them making music together even in winter when the river was so cold it made it hard for me to move."

"What kind of music did they play?"

"This was a long time back, Monk. There weren't so many fancy names for everything then, and the music wasn't as angry as it is now. Music isn't meant for anger. Wrong tool for the job, like me trying to bite that sturgeon with my tail."

"What sturgeon?"

"Are you blind? How can you miss a fish that big?"

I glance where he gestures with his chin, and sure enough there *is* a big sturgeon there, an impressive-looking river fish two meters long with white lateral markings, whiskers, and a shovel-shaped snout. The sturgeon's eyes fix on me but without any evident comprehension. A moment later, it disappears into the murk.

"Okay, I see him," I say.

"We've been dancing all these years, that old boy and me. He came up the river from the Gulf before they dammed it, and now he's trapped here, away from the rest of his kind, and growing bigger and more bitter every day. He has wanted to eat me since I was a baby. He used to chase me clear across the river. I had to search up rocks and hollows and mudholes all the way across so he couldn't get to me. Oh, how he tried with that great digging nose of his! He would have lifted all the

WE CHOOSE THE GODS THAT SUIT US BEST

mud off the bottom of this river if he could have done it. Made my days and nights a living hell. All I did for my first ten years was survive him. Of course, if he *had* gotten me, if I was in his gut and dead, he would rue what he'd done because he'd find himself all alone. Now, maybe a century later, the claw is on the other leg, if you receive my meaning. He comes too close and he's mine. He could go to another part of the river. He's a fish, don't you know, but he stays around."

"I think he likes you," I say.

"I think he's just lonely. I would be too, if it weren't for the God of Music."

"So your god appeared to you because of how much you love music just the way you Turtle Immortals appear to me..."

"...because of how much you love turtles," the snapper finishes, bringing his forelimbs together in an approximation of clapping his hands. "Good for you, Monk."

"I didn't realize that in these transmissions I was supposed to figure things out for myself."

"Why wouldn't you? Active thinking is better than passive listening. You should know that from your physical practices—your tai chi and all that. If you have to work for something it means more to you, is more likely to have a lasting effect on you, more likely to change you."

"Now I just feel stupid. Whether we choose our gods or they choose us, it makes perfect sense they should be in a form and context that has meaning for us, that provides succor and solace, that eases our suffering."

The turtle nods. "See this fishhook in my beak? Never have been able to get rid of it. Hurts me every day. For a long time, I found it easier to deal with because of the God of Music, because I loved him and he loved me. He would come down to the dock, sometimes with his woman, sometimes with his kid, and play on and on for me. I'd float on the surface, not even worrying about the fishhook or the sturgeon or anything else. If I'd been eaten at any of those moments, I wouldn't have cared at all. It would have been a fine way to go. These days, when I'm not on a mission like this one, I'm with him all the time."

"I wish I could play some music for you," I say.

"Thanks for saying that, Monk. You're a nice fellow. I heard that about you."

"So that was it? The wife died and you never heard any more music?"

"I was here longer than he was, and he knew that would be the case. So, he took care of me. Send his kid afterwards. That's how I know about Elvis. There are speakers made for swimming pools, for people who swim laps half the day. The boy dropped one of those in the river so I could hear the tunes. He used a radio at first, then a CD player, then a little gizmo that streamed free music. Helped me with suffering worse than the fishhook. Of course, that boy's no young man anymore."

"What could be worse than a hook you can't get out of your mouth?"

The snapper let go a laugh, sending bubbles the size of large pizza pies to the surface. "One time, I went after the sturgeon and got a soda can instead. I didn't mean to swallow it. Didn't even know it was there. The can was all twisted up and its sharp edges cut me on the inside while they were passing through. I couldn't move my legs. I had to lie there and just wait for it to come out. All that time, I just focused on how much the God of Music loved me and all he was doing for me."

"He might have spared you the can," I say mildly.

"That's not how gods work. Not how immortals work, either. You should know that. Like I told you a minute ago, we're not here to do the work for you, we're here to help you understand why the work matters."

"The pollution in this river," I say. "The terrible things you've been through and seen."

"Those were all so I could share them with you," he answers. "This is part of your spirit-writing now. To understand about how our gods choose us and we choose them, to spread the word about what humans are doing to the world. I'm a top predator, but nature can still make me suffer. You're top predators, too, and the same thing happens to you. You have to care for each other, and the world…"

"The way you care for that sturgeon. Loving him even though—"

"Just like that. And you have to understand what your gods are for and pay special attention to them."

"I have to go," I say. "Apparently, the lesson is over. But I have one more question."

"Shoot."

"That key around the God of Music's neck. What did it unlock?"

"Really? You don't know?"

"I don't."

"My heart, of course."

He settles back down into the bottom of the river. That big sturgeon closes in, gills fluttering. He flashes his gleaming belly at that fearsome beak, zooms away with a flick of his tail, then circles back as the giant maw opens and the tantalizing, tricky tongue twists. A wizened old man trundles toward the river. He's got a rough-hewn cane in one hand and a bright yellow loudspeaker in the other. He walks all the way to the end of the rickety dock, and sends it down by rope, inch by inch, to the surface below.

"Here you go, old friend," he says, leaning back against the dock. "Here's a little Beethoven for you."

COMPASSION CAN
BE COMPLICATED

It's a quiet day in the park today, punctuated only by the rumbling of a cluster of thunderstorms hovering out east over the ocean. Some ground iguanas, imports that have escaped from a nearby reptile wholesaler, have colonized the area and now peek out of holes in trees and scamper across the ground as if their tails are on fire. Sometimes when I'm meditating, I can hear the sound they make as they pass over dried leaves. Perhaps because it's a peaceful day and I don't have to try so hard to get into a turtle trance, it takes a while for me to settle down. I guess it's because just being there in the park is relaxing enough.

When I finally do achieve a different state, I find myself hovering over a crowded, tropical landscape. Traffic on a wide road below goes in two primary directions but is largely helter-skelter. There are small trucks and smaller cars and thousands of motorbikes. The presence of tuk-tuks, three-wheelers with bench seating behind the driver, tells me I'm in Southeast Asia. There's a small lake nearby. I lower myself toward the water and find a group of beggars scattered about. They hold signs proclaiming, in English, that they are victims of Agent Orange. One of them has no legs. Another has no face—his eyes are missing, his ears are missing, there is a single hole where his nose should be, and his mouth is a thin, undersized line. A placard identifies the lake as the one airman John McCain parachuted into after his plane was shot down.

This is Hanoi, the capital of Vietnam.

I return to the road. Traffic has jammed up near the lake. A small red pickup truck with its tailgate down stops quite near the water. I see a Styrofoam cooler in the back and notice the lid is ajar and moving. The beautiful, yellow head of a female Vietnamese box turtle appears. She looks around, sees the lake, and begins to clamber out. Her eight-inch-long shell is domed and keeled, and a rich brown in color. Her claws gain a purchase on the Styrofoam, and she pushes off the lid of the cooler and drops to the bed of the pickup with a thud. She's on her back, displaying her short tail and a yellow plastron. She has some trouble flipping over, but eventually, uses the swell of the wheel well to set herself right.

"Hey," I say.

She looks around and finally sees me. "Monk," she nods. "How are you today?"

"I'm fine, thanks. What's happening to you. What are you doing in the back of that truck?"

"Teaching you a lesson about freedom, I suppose. And about compassion."

"Thank you. But who put you there?"

It takes her a few tries, but she manages to claw her way to the top of the wheel well, from which vantage point she can look around. "The driver caught me," she says. "I've been so good at hiding in that pond, but he found me at last. Before today, he took my mother and he took my brothers and he took my father and he took my sisters. He has been the bane of our family's existence for years. He and his friends have taken so many of us, I don't know if there are any left at all."

"Where does he take you?"

"I don't know. None of us has ever returned to say. I fear the worst."

"The soup pot?"

"I won't speak of it. I refuse to accept that my entire family has met such an end. I just want to get away from your kind."

"I'm different," I say. "I've no desire to harm any living thing. I avoid stepping on ants. I don't eat animals. I save spiders on my walls using teacups and saucers and put them outside. Okay, sometimes, I kill cockroaches, but I'm not happy about it and I'm not proud, either."

The box turtle steps up from the swell of the wheel to the edge of the truck. She's clearly weighing the height of the fall if she just jumps. She tears her eyes from the road long enough to give me an arch look. "I eat roaches," she says. "They are a species of exception. But there aren't so many around my pond anymore. My mother told me they used to be under every leaf and that they tasted delicious, but then, many years ago, the orange rain came, and they all died, along with every egg my mother laid for nearly thirty years. Now when I manage to catch and eat a roach, it tastes bitter."

"Agent Orange," I say.

"Do you know what I have to do now to get a decent meal outside the pond? I have to stomp the earth like my father taught me. Lightly, to imitate the rain and make the worms come up so I can grab them."

The truck starts to move again, but not before the turtle scampers off the back and falls to the asphalt. I cringe, expecting her to be immediately hit by traffic, but if drivers here are accustomed to anything it is obstacles, so they swerve around her. I find myself yelling at her as I might an errant child.

"Are you crazy?!"

"I'm going home pond by pond," she yells back.

Dragging her shell on the ground, she crawls with astonishing speed toward the lake. She makes it to the circumferential foot path and scampers past the beggars, aiming at the bulrushes by the pond.

"Almost there," she pants.

I've heard turtle pant before, sometimes when male tortoises fight, sometimes during courtship, sometimes when chasing prey, and sometimes when afraid and rushing for cover. The driver of the red pickup truck, a rangy man wearing ankle-length trousers and a straw hat, leaps from his vehicle, shouts, and sprints off in pursuit.

"Don't look back," I warn her. "He's coming."

I try and trip him, but I can't seem to have any effect on him. I don't even think he sees me. Apparently, a feature of this lesson is that I'm there for the turtle but not for anyone else. She makes it right to the edge of the water and desperately shoves her forelimbs into the lake, but the man falls on her. A moment later he stands, both his shirt and hat soaked, and raises her victoriously aloft.

"I'm so sorry," I say. "I tried to stop him."

"I'll escape however many times it takes," she says grimly. "This isn't over yet."

The man runs back to the truck and tosses her into the cooler. He restores the lid, closes the tailgate, gets back behind the wheel, and drives off.

"Are you all right in there?"

63

"No, I'm not all right in here. I have no idea where I'm going and I can't see anything and he keeps cold beer in here, so I'm freezing."

"Who is this guy?"

"A hunter after the most lucrative prey. Turtles are big money, endangered species like mine all the more so."

"This isn't a third world country anymore. There should be better ways for him to earn a living than catching turtles."

"He found our pond. He's an opportunist. You all are."

"I think that's painting with a pretty broad brush," I say mildly. "You can go with the flow and still have a moral compass."

"We turtles don't bother layering complex ideas onto the realities of daily life. We think it's better just to live. Your kind is obsessed with accuracy, with measuring rather than just experiencing things. Each of us is fluid, ever-changing, and intensely unique. Why isn't that enough for you? Why must you always take one step forward and two steps back? Why must you drag your feet and complain so much?"

"The turtles I talk to all seem pretty down on the human race," I say.

"For good reason, wouldn't you agree?"

The truck traces a Byzantine route through the city. The pursuit finishes at a dead-end alley up in the central business district. The driver parks in front of a small, one-story, concrete building surrounded by a barbed-wire fence. I notice security cameras on all corners and by the only gate. The driver retrieves the turtle, now pulled fearfully back into her shell, and enters the facilities of a dealer in exotic animals.

I follow. The first thing that strikes me is the stench, which is part tangy urine, part pungent feces, and part coconut-fragranced disinfectant. Then there is the noise and movement. Monkeys scream, birds shriek, frogs croak, geckos chirp, fish jump, pangolins scratch, snakes slither, lizards climb atop their own dead, garishly colored newts paddle through fetid water, exotic rodents quiver inside paper towel rolls, and countless turtles splash.

The man in the wet straw hat exchanges the escaped box turtle for a thin wad of cash. Now she is in the care of the animal dealer, a diminutive man with a damaged eye riding low enough on his face to give him an unbalanced look. Bad Eye takes her to a room behind his office, places her on a metal table, pulls up a chair, and waits for her to come out of her shell. When she doesn't, he leaves the room.

"Is it safe to come out?" she asks me.

"For a moment, yes."

"Where am I?"

"You don't know? You're an immortal."

"Sometimes your transmissions are in the form of telling, sometimes in the form of showing. I'm living this story right along with you."

"In that case, I have to tell you I think you're in a really, really bad place."

"Do you think it might be the same place my family ended up?"

"I can't say for certain, but I am afraid it might be."

"And what kind of a place is it?"

I describe what I've seen while she has been in her shell and tell her what little I know about the trade of exotic species of animals. I speculate that she and the other creatures in this terrible building might end up anywhere in the world. I don't sugarcoat what often becomes of victims so treated.

As if on cue, Bad Eye returns to the room in the company of a painfully thin and elegantly dressed woman in a Western business suit and high heels. The flower-pattern scarf at her throat is by a luxury Italian brand. She and Bad Eye converse in Mandarin, by which I divine she is high-class and Chinese. She picks up the box turtle and turns her over in her hand, assessing her weight. She pushes against the turtle's rear leg with a finger and appears satisfied with the strength of the response. She hands over a wad of cash far thicker than the one Bad Eye gave the man with the straw hat. Bad Eye shakes his head. She adds a bit to the wad. Bad Eye shakes his head again. She adds a bit more. Bad Eye shakes his head and reminds her of the turtle's exquisite rarity. She stares at the turtle for a moment, then withdraws all the money from the table and puts it back into her purse. She stands and goes for the exit.

Bad Eye follows her. Outside, he puts his hand on her shoulder and she turns and looks at the hand in distaste. He says okay and she gives him the money and takes the turtle. A taxi is waiting for her. She goes into it and so do I. There is a small suitcase on the seat. She opens it. The driver glances in the mirror and she tells him to put his eyes on the road. Inside the suitcase is a bronze turtle. She fiddles with it and it opens with a cleverly hidden hinge. It has a thin lining of lead inside. She places the real turtle on the lower half and closes the upper half to secret it away.

"I can't see anything," she says.

I explain where she is.

"Why would someone put a real turtle inside a fake one?" she wants to know as I describe her prison for her.

"The lead gives me an idea, but we're going to have to wait and see to be sure."

"I've seen plenty of fake turtles, but usually they're made of plastic," she tells me. "All turtles know plastic very well. Plastic bubbles live in my pond. Little crabs eat them. Little shrimp, too. And I've seen plastic bottles and I've seen plastic bags. One of my brothers got caught in a bag and we tried biting it off him, but it was too tough, even for four of us at one time. We watched him die, pushing himself around with that thing on his head for weeks. The worst of it was, he couldn't eat and he couldn't drink. Nobody could believe he suffered for so long. I can still hear his moans in my head."

"If aliens ever come here from another world, they will probably think that the purpose of humans on Planet Earth is to take the remains of dead animals from the past, refine the sludge of them, and turn them into plastic," I tell her. "Really, this big blue rock in space is just a plastic factory."

As I suspected it would, the taxi heads for Hanoi's Tan Son Nhat International Airport, and as it was with the hunter, the driver cannot see me; my presence is known only to the turtle. Once there, the Chinese woman checks in for a flight to Hong Kong. She heads for security. The x-ray screener tells her to open her bag. She does. He takes out the bronze turtle and examines it carefully. She taps her foot expectantly and tells him it's a gift for a friend. He nods and puts it back into her suitcase.

"What's happening?" the turtle asks.

"Did your mother and father explain warplanes to you? From the bad times?"

"Bombs," she says immediately. It is the first time I hear fear in her voice.

"Yes, but not in this case. Not all planes are for war. Some are just to move people around. Sometimes we fly to see the people we love, sometimes we fly just to see new countries or to do business or to escape persecution."

"I just need to get back to my pond."

"I understand," I say. "Anyway, you are about to get on a plane. You may feel a great sensation of speed, but the flying itself is quite safe."

"This flying business is just a symbol of impatience," she says. "You know that the biggest problem your kind have is that you feel you have to keep moving."

"The human body was built to move."

"Indeed, you have evolved to experience the world at a walker's pace, not a jet pilot's pace or a motorcycle racer's pace or even at jogging speed. Anyway, I don't mean move to find mates or find food or run from a tiger or move to strange, rhythmic sounds, which I have seen your kind do. I mean move for no reason. I mean move because you are driven by inner disquiet. If I had to guess, I'd say the problems of the whole world arise from your kind's inability to stand quietly, without moving, for more than a few seconds."

"Most people don't care to, but I stand still daily," I say. "It's a quintessential Daoist practice and what led to me being able to see you."

"I hope your standing still cures you better than the machines I've seen your kind carrying in their hands, always staring at them with eyes wide, looking for answers but finding none."

The Chinese woman boards the flight and so do I. Perhaps because the cold cabin leaves her torpid, the box turtle doesn't say one word to me during the two-and-a-half-hour flight. When the plane lands, the woman goes through immigration and leaves the terminal. A driver in a Mercedes-Benz collects her at the curb and she is whisked away to a palatial estate atop Victoria Peak, the posh location of some of the world's most expensive real estate and a Hong Kong neighborhood I have watched become quite crowded and commercial over the years.

"The woman who bought you is about to take you inside a beautiful mansion. Her home, I think. I bet it has fantastic views of the city."

"Is there a pond nearby?" the box turtle asks hopefully. "Do you see any of my family members? When will she take me out of this box? It's hot and hard to breathe."

"Soon, I'm sure."

The Chinese lady leaves the turtle on the entryway table and walks to her bedroom. I follow. She unpacks her suitcase then passes through a spacious living room to a garden with a salubrious view of the city and harbor beyond. There is a pond here, in grand *feng shui* style. The fins of carp and the heads of turtles protrude from the water. There are also various box turtles on the basking rocks and on a narrow swale of leafy plants. These include Vietnamese varieties like my friend from Hanoi, along with rare ones from Hainan Island and from the tropical, southwest Chinese province of Yunnan. I harbor some small optimism that some of these turtles are my new friend's lost relatives and look forward to a possible reunion. I try and talk to the turtles, but find I cannot. The lady looks at them with satisfaction, and returns to her bedroom, where she removes her makeup and her dress and collapses upon the bed.

Feeling the reluctant voyeur, I see she is not merely thin, but emaciated, her protruding ribs and sunken cheeks making that clear. Her ribs protrude, and gaunt, sunken cheeks confirm what I have suspected—she has cancer and it is advanced. After a short nap, she goes out to her turtle pond. She selects one of the turtles there, a male box turtle of the same species as the immortal still incarcerated in the smuggling container. She brings the turtle into the kitchen.

A housemaid appears. The maid runs the faucet until it is steaming hot, then drops the turtle into the water. It comes out of its shell at once, desperately paddling to be free of the scalding bath. The maid produces a pair of poultry shears and summarily snips off its head. The eyes bulge in utter disbelief. The jaws work frantically. The cervical spine, white and twitching and dripping threadlike nerves, searches desperately for the body. The maid takes a few cuts of the meat of its forelimbs while it is still alive and sets them aside, bloody, on the counter. I have the impression that this meat will be used in soup later.

The maid turns off the tap and throws the turtle's head into a trashcan under the sink. She takes a small saw from a drawer and uses it to separate the carapace and plastron. She scoops out the organs and cuts away the limbs. She pounds the shell with a mallet, reducing it to flat, wet chips the size of casino tokens. She wraps these chips in a cloth and pounds them some more. Once she has smashed the shell to small pieces, she puts them on a cookie sheet, places them in the central rack of the oven, and sets the oven to bake.

All the while her employer leans weakly against the counter, watching the proceedings, her energy apparently drained by her trip to Vietnam. Once the baking process is underway, she opens a cupboard door. Inside, I see row after row of tightly fastened jars, each brimming with blue-black, homemade turtle jelly, a combination of herbs and ground-up turtle shell, a traditional folk cure for everything from irritated skin and poor appetite to cancer. She opens one of the jars and smacks her lips as she consumes it, exhaling loudly after each spoonful. When she has had enough, she returns to the entryway, removes the Vietnamese box turtle from its lead case, and carries it out to the pond out back.

"Where am I now? What's happening?" the box turtle wants to know.

I'm not sure how to play this. I'm not sure how much the immortal knows and how much she doesn't. Each of these transmissions is there for me to figure out, and I decide just to follow my compassionate instincts on this one.

"You're in your new home," I say. "The place where your longing and pining and waiting is over. The end of your suffering. The place where all your dreams come true."

"Oh, thank you," she says sweetly.

A moment later, she sees the pond. She sees the clean, clear, slowly moving water, filtered and pumped to perfection. She sees the luxuriant shade of the philodendron plants and she sees the perfectly placed flat basking rock. She sees the heads of her brothers and sisters and the shell of her mother, too. She wriggles wildly in the Chinese lady's hands until she is finally set free at water's edge.

She swims immediately to her mother. Together, they emerge from the water and stand on all fours, necks extended, nose to nose.

"Where's father?" she breathes excitedly.

Her mother blinks. "I'm not sure. He was here a few minutes ago," she says.

A moment later, I'm back in the park. The thunderclouds are gone, but the air is still heavy.

WE HAVE TO BE TAUGHT TO HATE, WASTE, AND DESTROY

Unpredictable things often happen during the prelude to meditation. The body can resist the discipline of the practice, evidencing shivers or shudders or pain. The mind can resist, too, excavating the past only to come up with longing, nostalgia, or regret. It can generate anxiety about a future we can neither accurately predict nor control, humoring compulsions or obsessions or titillating itself with fascinating ideas.

Today in the park, my mind likes that last option, and I find myself thinking about the power of myth, not only in spirit-writing but also in the popular imagination. Great scholars have devoted careers to the importance of myths and legends, documenting how they changed the course of society, flavored culture, created archetypes that set the tone for countless individual lives, and, sadly, prompted wars.

Studying mythology yields deep and often fascinating information about cultures long gone—archeologists and anthropologists often extrapolate written or painted stories into conclusions about the way people lived and what they believed—but also reveals a lot about the modern world and contemporary culture. If we were fish, we might say that understanding mythology shows us the water in which we are swimming but otherwise would not notice at all. I drift into my immortal trance thinking about dragons that are really just alligators with wings, gods wielding thunderbolts, dwarves with axes, elves with swords, wizards with wands, and, inexplicably, sumo wrestlers carrying plates of enchanted sushi.

I come awake at dusk, lying on a bed of humus, leaf litter soft against my cheek. I rise into cool, foggy air and take a few dizzy steps, steadying myself against the trunk of a birch tree. The forest is dark, moody, and invigorating, and I know at once I'm in the kind of special place where the environment is an important part of the lesson I am about to learn. A deer moves past me, stops to stare, nose twitching, ears erect. A raven alights on a branch level before me and I smell its musky perfume.

"Good evening," he says. "I'll need you to follow me, please. Chop chop. No time to waste. The light is going, and the immortal wants to talk to you before dark."

It is a measure of how accepting of these magical trances I have become that I follow the bird without question through a disorienting forest kaleidoscope of birch, elm, lime, and oak trees with thick trunks and delicate canopies. Vines impede my progress and tangled roots restrict my steps. Creatures flicker in and out of view. The ground beneath me rises and falls. A man in a cloak rushes past, panting heavily, brown robes tied tight around his ample belly. He's followed a moment later by a giant with an oak staff in his hands. The wood gleams from the polishing effect of his calluses. Neither man pays me any mind. I plop myself down on a stump for a breather. The raven hovers above me. A bug the color of a burned apple explores the edge of my monk slipper.

"The hazel pot beetle," says the raven. "A great treat. I used to eat them daily. So rare now, I dare not. Have to do my part for conservation even if it means going hungry."

"What a strange little fellow," I say. "Can't even see its head, really."

"Hidden in the thorax. *Cryptocephalus*, in Latin, means hidden head."

WE HAVE TO BE TAUGHT TO HATE, WASTE, AND DESTROY

"You speak Latin?"

"We're in Britain. Where else would you expect to find a
Latin scholar?"

A third man steps out of the trees. He's handsome in green tights, a
hood, and a goatee and mustache, a longbow in his hands. He grins at
me, then melts away into the woods.

"Okay," I say. "Enough of this. I'm not a child."

"Don't you *want* to be a child again? Isn't that what your most vaunted
sage, your Laozi suggests? To be able to experience the wonder
of nature and find joy in life's unexpected and inexplicable twists
and turns?"

"It is," I say.

"Well then. You are in Sherwood Forest, the epicenter of all that
is mystical."

"You're joking, right? Are you going to tell me I've just seen Friar Tuck
and Little John and Robin Hood himself?"

"You have."

"This is kind of silly," I say. "Like mixing metaphors. I'm Daoist."

"When history, mystery, memory, and magic frolic in the bosom of
nature, they transcend culture and tradition," the raven says. "You will
find there is much to be learned here."

Someone blows a horn. An arrow whizzes by my head. I duck and run.
In my haste, I stumble onto a piece of raised ground extending about
four meters in all directions. There are stones in it, fallen steles maybe,
or just markers of some kind.

"You may not realize it, but you just stumbled over history," the raven observes. "There are several important places in this wood, including lodges from when it was a royal hunting preserve. The Vikings called this mound *Thnghowe*. People has used it for rituals for at least four thousand years. The group of three stones marks the parish boundary. The one in the middle was the foundation stone. It was a gathering place for politics in your so-called Middle Ages. Rulers of antiquity may have been buried here, or nearby."

"How come you know so much?"

"Every raven's a history maven."

I follow him through a double row of lime trees to a ramshackle ruin, gloomy but beautiful due to the multihued bark and lichen-rich stones glowing in the late day sun, redolent of a deciduous forest in summer, with aromas floral, earthy, and pleasant. I close my eyes and inhale the fragrance. When I open them again, a female marginated tortoise, largest of the European clan, has materialized at my feet. She stands on her rear legs, her neck outstretched into the last rays of the fading sun. Her black-and-yellow shell gleams.

"Felicitations, Monk," she says.

"And to you, Immortal."

"The raven's right about this place you know," she says. "Nottinghamshire was once the Garden of Eden. That's why I've stayed, and why the royals took it for their own. Hunting was just an excuse. The Celts, the Vikings, and especially the Druids all knew this to be an energy vortex. Have you been to one before?"

"An energy vortex?" I ask. "Well, I went to Sedona, Arizona, a few years ago. They say there are several there. And Mauna Kea on the Big Island of Hawaii."

WE HAVE TO BE TAUGHT TO HATE, WASTE, AND DESTROY

"Such places birth legends just as surely as oppression leads to rebellion. The royals were greedy. They still are. They hired sheriffs to impose their will upon the peasants. Taxed, the peasants starved. Starved, they had two options: either they poached deer from the royal estates and cooked the deer to feed their children, or they became bandits and repossessed the taxes taken from them. A good bandit, one who shared what he retrieved, well, that was good fodder for fame. By the way, did you know that Sherwood Forest was later torn apart for coal mining?"

"I did not."

The immortal tortoise nods knowing. "Oh yes, all those poachers, peasants, and bandits went straight down into the bowels of the Earth on missions of manliness and spirit, mistaking plundering for discovering what it really means to be close to Mother Nature. They stripped the place bare and turned the Garden of Eden into a wasteland. The scars are still upon the land. They logged every tree, too, depriving countless creatures—"

"Mostly birds," interrupts the raven.

"—such as squirrels, of places to live," the immortal continues, undaunted. "If there was a way to destroy the land, so-called industrialists found it and did it."

The sun is moving, and the shade is too cool to keep a tortoise active, so we move a few paces in pursuit of warming rays. I ask the immortal how a tortoise from the sunny Mediterranean can adapt to such a cold place.

"We turtles are more adaptable than people think, although I credit the special magic of this place with sustaining me. Before these buildings were an estate, they were a monastery. You can probably feel the remains of devotional energy. In the old days, this place was for those who toiled, fought, and prayed."

"It's a big park now," says the raven. "There's a bike path beyond those trees. If you want to see some hot-looking women riding and jogging, I can take you there."

"Always the libertine," the immortal shakes her head half in amusement and half in disgust. "That crow's a Daoist without even knowing it, though this forest really is the epicenter of Western shamanism. There used to be Druids here."

"Is that why we're here? To remind me of the universality of shamanistic ideas?"

"You catch on quickly, Monk."

"Whereas Robin Hood was the village idiot," says the raven. "And his girlfriend was not at all the beauty they say she was in legends. She had a wart the size of a gooseberry on her cheek."

"The British government hid tanks and ammunition here during the war," says the immortal. "When you people had done with killing each other, most of the wilds were lost."

"What's left is a mere shadow," says the raven. "So much natural cover gone, so much lost and destroyed."

"I left my family behind when the royals took me," the tortoise continues. "I was barely a hatchling. Since then, it hasn't been pretty. Greece, I hear, is not the place I remember. Overrun with people, bridges, houses, parks, factories, warehouses. Thousands of turtles being killed by cars and tour buses. Tour buses! When I was born, there was no such thing. Do you see the irony of tourists killing the very land they've come to admire?"

"I see it," I say.

"Of course, thousands more of us were taken for garden pets and left to die under lights in the garage of so-called turtle lovers. If they loved us so much, I warrant they'd leave us be. I suppose you can see why I appreciate this place."

"But don't you get lonely being the only one of your kind here?"

"I talk to the raven."

"Never shuts up," says the raven.

"See how he is?" the immortal laughs. "Always has to get in a word."

"Believe me when I tell you she's the queen of mindless twaddle, Monk. Go ahead and ask her about the children—the dukes-and-duchesses-to-be."

"People say that human children are cruel," says the immortal. "I disagree. I think they're beautiful and pure. They have to learn to do the terrible things they do to each other and the world. I believe in basic human goodness."

"Perfect twaddle," says the raven. "All people are bad."

I suddenly realize something.

"You're an immortal too, aren't you?" I ask the bird.

"I would have to be, wouldn't I? To be here in this lesson with you?"

"I remember little Reginald with the gimp leg," the immortal tortoise says wistfully. "He would kick me, but never very hard. And Geraldine, who wore braces on her teeth, she liked to chew on twigs and was kind to me and brought me spring wildflowers. Then there was tall, strong Michael, who would blow on me with his hot breath because he knew I was cold all the time, and Penelope, who managed to figure out that I missed the pines of Greece and would pour retsina into a dog bowl and let me sniff and swallow it until I shit out what looked like olives. And Bruce, with white hair on his arms—he carried a badminton birdie around in his pocket. I can't forget Jennifer, either, who cheated at croquet and who never, ever laughed, nor little Pasha, who was swarthy and had a Persian name, which made me think his mother had strayed. And in the last generation, before all hell broke loose and the sky fell and the bombs boomed and the forest burned, one last girl, Guinevere, who was named for Arthur's bride because in those terrible days all everyone wanted was a reminder of the time when character mattered and men were brave because they wanted to be, not because they had no choice."

"She really loved those children," says the raven, uncharacteristically quietly. "She had none of her own, you see. Not in any of her lives. It created a certain bias…"

"The generations grew smaller each time," the tortoise continues, "but the quality of their thinking grew. That's the thing with you humans. As you get closer and closer to annihilating yourselves, you slowly wake up…"

"Your kind is totally rethinking the question of animal consciousness, for instance," says the raven. "Belatedly, but still."

"The question is, which side of you will win? The waking side or the suicidal side?" says the tortoise. "Did I tell you that before the bombs fell, there used to be a white picket fence around the garden?"

"Who cares about the garden now?" says the raven.

"In the early days, a special monk did," the immortal muses. "He was the one who told me about the special energy of this place. Even though he wore a frock and cross, he confessed to me he was secretly a Druid. He was old enough to have known the Great King, the Great Magician, and the Lady of the Lake. When the monastery became an estate, he stayed on as a gardener. None of the royals knew who he was, knew what he had achieved in the training of his mind, in the taming of his heart. They treated him worse than a dog, beating him and turning him out into the cold when even the hounds had a spot

79

by the fire. He slept in the barn with the horses and sometimes would suffer their hooves in thunderstorms. Even so, he helped more than one mare with a difficult foaling. Sometimes, the hunt master got drunk and beat him. The thing is, he could have wandered off anytime and nobody would have complained or even noticed. He didn't, though. That's how much he loved the land."

"Next you're going to tell him how he polished King Arthur's sword, Excalibur," the raven scoffs.

"Laugh all you wish, but you know he did."

"What happened to him?" I ask. "The monk, I mean."

"One day, we went out together. We walked slowly, because I'm small and he was old. He protected me from racoons and weasels, badgers too, but never by hurting anyone. We went to *Thynghowe*..."

"Where you stopped to rest," the raven reminds me.

"The old man rested there, too," says the turtle. "Sat very still for a long time. Finally, his breathing went quiet and his heart stopped. He remained warm even so, and I stayed pressed against him. Day after day, there was no smell and there was no decay and no vulture dared at him. It took the royals two weeks to find us, and then only because they were out hunting and penned a fox with their dogs nearby. He looked so fresh that they spoke to him and were annoyed when he didn't answer. The hunt master even hit him with a whip. Only when the whip drew no blood did they come in for a closer look. They took him for a warlock and burned his body at dawn."

"I remember that day," says the raven. "His pyre-smoke was as dark and black as burning wolf dung."

"When the fire was over, there was no trace of him," says the turtle. "There was no dust and there was no bone. Not a strand of his white hair."

"Pride's not my thing, but this story makes me proud to be a monk," I say.

"Should make you proud to be human," says the turtle. "It's the fact that I knew him that gives me hope for us all."

"Him and the children," the raven reminds her. "You loved the children, too."

"The monk and the children," the tortoise says, and as she does, the sun fails her, and the tree shadows come up, and my own shadow begins to tear at the edges. I watch it melt into the grass until, at last, Sherwood Forest is gone and I'm back in my mediation park.

It feels a disappointingly regular place today.

❖ WE HAVE TO BE TAUGHT TO HATE, WASTE, AND DESTROY ❖

INVESTING IN LOSS

The park is quiet this morning. As I regulate and relax my body and prepare for a long meditation session, I hear a curious combination of wheel noise, huffing, and little squeals of delight. Through squinted eyes, I see a tow-headed boy in blue shorts and a yellow shirt riding a large-wheeled red bicycle that seems too large for him, with the seat and handlebars set at their lowest position. A bald, middle-aged man trots behind him, issuing encouraging words, breathless and bent over, one hand trailing, another firmly holding the bicycle seat so the boy stays upright. The boy pedals faster and faster, careening down the path, until the man has no choice but to let go. He hunches over, gasping for breath, as the boy squeals with delight.

"Look Grandpa, look!" he cries. "I'm riding by myself!"

A brace of ducks comes up out of the water right into the boy's path. He veers off so as not to hit them and ends up in the grass. Amazingly, he doesn't fall, but continues to pedal, more slowly because of the increased rolling resistance at the wheels, until finally he comes to a stop. He sticks his foot down and half topples, half dismounts, punching his fist triumphantly as his grandfather trots over, arms wide, smiling broadly.

The scene reminds me of when my own grandfather taught me to ride, also on a red bicycle a bit too big for me. Reminiscing, I find myself a bit teary-eyes and lost in memories. These unfold rapidly, like tapping hyperlinks on a computer, and I get so lost in them and their directions that it takes me a moment to notice I am no longer in the park but rather flying over mountains so tremendous their peaks scrape the sky. I drop down into a valley and follow the course of a river running fast with snowmelt until I come to a warm, green lowland. The river narrows and slows besides a tropical pasture, the white-topped curtain of mountains providing a magnificent backdrop. In the river shallows, three men sit in a coracle, one at a guiding pole, one at nets, and the third standing with a thick fishing rod in his rough hands. The line is wire, not filament, and it is attached to a car battery.

It takes me a moment to recognize what is strange about the scene. While the boat is not exactly a sexy America's Cup racer, it does have a bow and stern. Somehow, however, the stern is leading the way. As I get closer, I see the fisherman is braced against the gunwale, his rod bent fully horizontal. I follow the line to the surface of the water and then, because I can, I enter the river.

It is hard to see the fishing line, but keeping the angle of entry in mind, I project its course down to the bottom, just a few yards away. The wire leads into a harpoon whose triangular tip is buried in the flattened carapace of an enormous turtle. The way the animal jerks spasmodically tells me it has not only been pierced but is being electrocuted. The shape of the head, the four-foot-long leathery shell, and the pattern of variegated yellow lines on the skin all tell me this a male Asian narrow-headed giant softshell. I remember that these oversized freshwater turtles lay clutches as large as one hundred eggs, that they eat mostly fish, and that they are almost never seen because they spend most of their lives hiding under the sand or mud of South Asian waterways.

"Can you hear me?" I ask.

"Ah. Monk," he says, surprisingly calmly. "I've been waiting for you."

"How can I help?"

"If you touch me or the wire, you'll receive the current, too. That could stop your heart. I've got this."

"It doesn't really look like you do. How badly are you hit?"

"There is pain in a bone beneath my shell, but my spine and internal organs are safe."

"Thank goodness."

"There was no goodness at play here. These animals want to catch and kill and eat me. My kind is a delicacy here and abroad. They've been trying for years, though not with an electrified harpoon."

His thick legs are powerfully muscled. Even with the harpoon in him, the current making him twitch, and the weight of the three-man coracle dragging behind him, he makes progress through the water. Looking up, I can see the stern of the boat dipping down. After a few more minutes of pulling, he reaches a rock formation on the river bottom. He enters it and uses the edge of the rock to relieve the pressure of the line.

"My last lesson was sad and this one is, too. What am I supposed to learn from this?" I ask, feeling helpless and thinking this wasn't exactly what I signed on for when I agreed to these spirit-writing sessions.

"This time," the immortal tells me, "you have only to bear witness."

"Then I want to get all the details right. Where are we, exactly?"

"In a river."

"I don't recognize it."

I swim down to him and grab hold of the rock. He looks at me with a piercing look in his yellow eyes.

"You don't recognize the Himalayas?"

"I guess I should have since I flew in over their peaks. Are we in Tibet?"

"Certainly not."

"India? Uttar Pradesh?"

"Too many people. I would never live there. Also, my job is here."

"Nepal, then."

"Bravo."

"And what is your job?"

"A monk like you should know the answer."

"Monks in Nepal are Buddhist. I'm Daoist."

The immortal fluffs the water with his front legs, stirring up the sand and obscuring him from view until it settles again.

"I knew that," he says. "I'm just a bit busy to think clearly right now. And, truth be told, most of the monks I've met have more of a presence. You're a bit ordinary-looking."

"Being nameless and low profile is a Daoist ideal," I say, "although, honestly, some people tell me I look like a turtle, the way my nose sticks out and my head juts forward on my neck."

"What a magnificent compliment."

Despite the electrical convulsions, the big softshell is very precisely rubbing the harpoon against the rock ledge beneath which he hides. Softshell turtles, as their name suggests and in some ways like the leatherback who spoke to me from the Pacific Gyre, lack the fingernail-like keratinous shell layer of most turtles, having instead a leathery covering over bony plates. I go in for a closer look, notice the puckered, red-and-white tissue erupting around the entrance wound and that the harpoon is not so deeply buried as it was when I first saw it.

"You're very clever," I say. "Using the rock to dislodge the harpoon."

"That seems clever to you, Monk? What else would I do? Even an ant can rub against a tree to dislodge a speck of dust. You know what I *don't* think is clever? Defining intelligence as something familiar and recognizable the way you humans do, and then linking intelligence to value."

"I was just making conversation," I say. "Since I can't help and am only here to bear witness and all. If there are so many hostile fishermen here, why don't you swim somewhere else?"

"My family has been guarding this sacred place for thousands of years."

"What's so sacred about it?"

Before the softshell can answer, the fisherman from the coracle jumps into the water wearing a dive mask and holding a large knife. He follows the wire toward the harpoon, all the while careful not to touch it. In response, the immortal turtle works the wire against the rock with new energy until it finally goes slack. Free, he rises like some primordial creature of Daoist legend, a pure, unapologetic manifestation of nature in all her glory. The diver pursues him and receives for his trouble a vicious bite to the Achilles tendon and another to the back of the neck. Screaming, the man surfaces just in time to see the pole wrenched from his companion's hands, bringing another victim into the water. The turtle turns his attention to the boat itself next, capsizing it with a ramming attack and sending the third fisherman into the water, too. As all three men make a cursing bid for the reedy riverbank, cell phones, shoes, Buddhist prayer wheels, and outwear for the brisk Nepal nights sink slowly through the water column.

I've seen bigger, stronger animals evidence great feats of endurance and tenacity after being wounded, but I've never seen any nonhuman creature apply such battle tactics against human beings. "Did you accept the harpoon willingly?" I ask him in a flash of insight. "As a way to physically connect and affect your adversaries?"

"Sometimes we must lose to win," he says.

"You planned this. You let them strike you, but took the harpoon where it did the least damage. You feigned weakness to encourage them to come after you."

"I feigned blindness and pretended I didn't see them."

"You let them think they had you cornered, let them think they had lured you from the deep waters of the center of the river to the shallows, when, in fact, the shallows were exactly where you wanted to be. You headed straight for those rocks. You knew you could use that overhang to free yourself from the harpoon."

"As you say."

"Deception. A core military principle put forth thousands of years ago by Sunzi, a sage general in my tradition who penned a classic called *The Art of War*."

"I may be playing the role of a Buddhist, but I am a Daoist immortal, Monk. Do you really think you have to tell me about Sunzi and his *Art of War*? We immortals have been employing such principles for generations in defense of this river. We do, after all, have things to do besides tutoring you."

"Of course," I mutter. "Silly of me."

Hearing shouting, I turn to see activity on the riverbank. Upon seeing their sodden compatriots climb ashore, a group of fishermen begin gathering. They gesticulate, and I sense they are making plans.

"I don't think this is over," I say.

Just as I make this pronouncement, a bale of giant turtles swim in and darken the river. They close ranks around my teacher.

"One more thing before my final battle begins," he says.

"Yes?" I manage, amazed by the number of turtles, by their size and ferocity, by their resolute expressions and precise formation.

"Violence is the lowest common denominator of life on Earth. There's no point in fighting if you don't know what you're fighting for. You must identify the plan heaven has for you—"

"*Ming*," I interrupt.

"Just so. *Ming*. In your case, it is spirit-writing that will move others of your kind to stand against the multiple transgressions being committed against Mother Nature."

"Yes," I say.

"To do this, you must build your strength of will—"

"*Zhi*," I interrupt again.

"Exactly, and use that to bring your essence, your energy, and your spirit to bear on the battle," he finishes. "Find and share love for the entire sentient world, Monk, as it will make this work easier. Gather yourself and your people. There is much to be done."

As these last words come to me, I see an armada of coracles putting into the river, each laden with armed fisherman. I see harpoons aplenty, and nets and knives. I hear angry shouts and I hear prayers for victory.

As the massive regiment of giant softshells assumes a phalanx formation and prepares for war, I find myself back in my park.

WE'RE ALL IN THIS TOGETHER

I'm meditating in a rare southern cold front today, with a brisk wind that roars in the trees and challenges me to find an angle in the copse wherein I am not blown about. I take little side steps from my standing position, touching the ground with my heel first. The wind is a devious adversary, shifting me just enough, first one way and then the other, to make it hard to find stillness, either in the wind or in my mind. There are special proscriptions in Daoist practice against baring either the throat area or the low back to a chilling wind—particularly if one is in the same position for an extended period of time—and I take them seriously. After a time, I find a place, much closer to a big oak than I usually get. At this distance, I run the risk of birds making deposits on my head, lizards crawling onto my shoulders, or spiders investigating my orifices. All the same I have no choice. Snug against the tree, the gale is less tiresome.

I drift and drift until I feel like a tree myself, growing upward from the earth. The sky is vast, and the smell of springtime dirt grows stronger until it's clear I am no longer in my meditation park. I'm in a valley at the base of a craggy brown hill eased by a stubble of green grass. In the background is a vast blue sky graced with a cluster of thin clouds. Closer, the hill has been excavated on a large scale.

Inside what could be a church alcove, if it weren't a hundred feet high, are the remains of a giant statue, giant feet with ruptured toes, a fractured midriff draped in a robe, one lone finger, broken and without context, pointing downward.

Beside me is a large hole in the ground. Just as I see it, a massive explosion shakes the ground and a plume of dust rises from the valley floor. Terrified and seeking cover, I crawl into the hole. The air is cold inside and smells like fruit. Roots growing like stalactites prick my bald head and something pushes against my toes. I look down and see the hard, sharp carapace of an old female Central Asian tortoise, her shell cracked, her skin weathered.

"Hello," I say.

"What a pushy monk you are," the immortal turtle comments. "I was going to come out and greet you. You didn't have to violate my burrow."

"Sorry," I say sheepishly. "There was a tremendous noise and I was startled."

"American bombers after Taliban targets. They're at it every day."

"So, we're in Afghanistan."

"Of course. Did you see what's left of the big Buddha statue? Religious fanatics destroyed it. What a sight it was. What an expression he wore."

Moving closer to hear her better, I see that she wears quite the expression herself, her eyes clear and bright, the wrinkled skin around them adding depth and wisdom to her gaze.

"Do you have eggs in there?"

"How is that your business? Just because you're here for a lesson doesn't give you license to pry."

I want to laugh but something tells me not to. "I like the fragrance of your burrow," I say. "It smells like raspberry to me."

"It's spring and berries are everywhere," she answers. "Though your clumsy attempt to change the subject didn't fool me. Privacy has always been important, now more so than ever with the world so overcrowded."

"Again, I apologize for barging in."

"Let's go out now and take in the sights."

We do, and in a few minutes another airplane appears, thundering through the valley to launch another missile at some target I cannot see.

"That's a woman at the stick," the immortal tells me. "I can tell from her smooth flying. Men are more compulsive. They correct course more jerkily and can't relax. They're fuel hogs, too."

Now I can't help but chuckle. "A turtle immortal who talks piloting. Wow. Is that my lesson for today?"

"I've been around. Now watch what comes next. See the what's in the cleft of that ridge to the right of the statue?"

I do, and at first, I don't see anything. Then, squinting, I make out a glint that resolves into a large metal tube, the gun barrel of a camouflage-colored tank.

"M1A1 Abrams," the tortoise says with some satisfaction. "Lot of firepower, but those treads are hell on tortoise burrows."

"You mean they run over you?"

"If an Abrams tank ran over me, would we be having this conversation?"

"I meant other turtles."

"I lost two daughters to that same cursed vehicle," she answers distantly. "One died on its outbound trip, another when it was returning. The second was full of eggs. There's an American base behind that hill. Taliban prisoners are kept there, in the center of the compound. Sometimes, when a summer storm puts water in the air and sound travels farther, I can hear them screaming. All of this is in the name of religion. Because of one concocted belief or another."

"I can't think of examples of Daoists killing anyone because they didn't believe in Dao," I say.

"It's not about how you view things; it's about intolerance. Humans can't afford that anymore. You're all crammed together too tightly. As soon as someone does something stupid, it's like dropping a match on dry brush."

"Speaking of dry brush, tell me about the seasons here."

She takes a few more steps and extends her neck and limbs to catch the sun in the most quintessential of all turtle postures.

"The summer's the one that counts," she says, getting a dreamy look in her eye. "Fresh foliage comes up out of the dirt, the sun makes the air shimmer and sway, the horizon becomes a mirage at dusk, the washed-out sky turns deep blue, and the wildflowers bloom so hard, so long, and so bright, you'd think they're the ones that lit up the minds of the great thinkers of Asia. It's the nights that are best, though, for the slow-moving curtain of stars, empty but pregnant with infinite possibility, reminds all of us that we are a part of everything."

"You're a poet!" I breathe.

"You're surprised? I'm an immortal."

"A turtle poet just seems a paradox," I say. "I didn't intend any offense."

"But turtles are all *about* paradoxes. That's why you came to us for lessons. We are helpless, yet have endured for tens of millions of years; mostly silent, yet we have so very, very much to say; dense and heavy, yet we can fly through the water; primitive, yet specialized and advanced; low-lying, yet we soar into space."

"Did you just say you turtles soar into space?"

"Admittedly, not *all* of us," she says. "But I did."

"Wait. What?"

Just then, the rumble of jet engines comes again, growing from the south, first faintly, and then louder. I glance around, looking for some cover other than the tortoise burrow, not wanting to give offense again but not feeling so great about standing out in the open, either. I consider the alcove holding the remains of the statue, but the evidence doesn't point to that being particularly safe.

"Relax," says the immortal. "The pilots can't see you. Satellites can't see you. Drones can't see you."

I exhale with relief. "Right. This whole dance, the rules...."

"This is spirit-writing, not a dance, as if Afghanistan cares. Do you know how many kings and princes, sages and conquerors, mystics and hermits and monks—yes, monks—have stood looking at this valley? There's an energy here that sucks the souls of foreigners from their bodies and neutralizes the crushing material force of technology. It may be karma, something left over from the Mongols, or perhaps something spiritual, cleansing, and protective deposited by all those Buddhist sages meditating in their temples and caves."

"You were telling me about space."

"It was the 1968 Zond program," she says as another fighter jet appears, flying fast and low, changing direction often in a fashion I gather suggests a man in the cockpit. "That's what gave me the perspective I have about the divisiveness of religion and the need for humans to come together."

"My lesson," say.

"Your lesson. It was 1968. A Proton-K booster rocket with a Blok D second stage. The capsule was a Soyuz 7K-L1 modified into what they called 'Interplanetary probe number 5.' Unmanned, of course, because people didn't have the balls."

"Sorry, but I'm totally lost."

"There were two of us, actually. My mate and me. He's gone now. The radiation up there was fierce. It was terrible to watch him fall apart long after we returned, the scales on his forelegs dripping off him, his shell softening, his eyes going dull to the point where he couldn't see. I led him back to the burrow, but winter came, and while I was asleep and he was asleep, cosmic rays damaged the special molecules in our blood that keep it liquid so our hearts can pump it and we don't freeze solid. Can you imagine that I had to push him out in springtime, stiff and sightless, and not only out, but up to the top of that far-off ridge so the vultures would come to him and take him, piece by frozen piece?"

"You went into space in a Soviet capsule!"

"There were fruit flies with us and worms and test tubes of bacteria. Some plants, too. The capsule was cramped. At first, they put us in men's large-size socks, but then some biologist pointed out that we couldn't breathe freely if we couldn't pump our limbs in and out, so they cut us free and let us sit inside a compartment near a window. We traded the view for my mate's life, really, and perhaps mine, too, for when I die, it will be from what that radiation did to me."

"Tell me about the view."

"We went clear around the moon to the dark side, then back to the light and the view of our home, a blue drop in the blackness. The guidance system failed when we reentered Earth's atmosphere, so we couldn't skip gently back in like we were supposed to, bouncing off the hard edge of air and then back up and then down again, scrubbing speed each time. Instead, we came in hard and fast. Landed in the Indian Ocean. The Soviet Navy sent the *Borovichi* and the *Vasily Golovin* to retrieve us. Some of the bacteria test tubes broke. Who knows what bugs still survive in me?"

"I can't believe you've been to the moon."

"We didn't land. And it was so unimaginably cold in that capsule, it makes the worst Afghan winter feel like Tahiti. I saw Tahiti, by the way. On the way down. It's beautiful. We lost 10 percent of our body weight on the trip, which is strange when you think that turtles are vessels for sunlight, powered as we are by celestial warmth. Not in that capsule, I'll tell you. We weren't powered at all. So, I still can't figure out how we got so skinny, other than the effects of those terrible, terrible rays."

"And you got back here, to your burrow, how exactly?"

"Exactly in one of those merciless helicopters, with its guns and its bubble nose and its huge slicing blades and all those seats for troops who won't ever come home to see their loved ones again. At least *my* love died next to me, pressed against me, in the cold dark of winter. Those boys, your people, half of them died here, blasted out of the air by missiles or shot in ambushes. The *mujahideen* know these passes the way a goat knows her own kid."

"Why were you in a helicopter?"

"Some Soviet scientist had a heart. She was just a kid herself. She had red hair. She thought bringing us back home was something she should do, not just home to Earth but here to the same burrow she took us from, carefully marked with academic precision by a little red hammer-and-sickle flag in the dirt. I still remember the way we three sat together in the middle seat in the middle row of that otherwise empty giant aircraft. She looked so small and frail, but she had a will, that one. She was and is my favorite of all your kind, even though after she kissed our shells and left us, I never saw her again."

"So you have an astronaut's perspective."

"Oh yes. And if I hadn't hurtled through the cosmos untethered the way I did, I would surely not have seen how great the overall balance is and how inexorably and unstoppably the universe corrects for whatever you people do. I lost my anger at your actions when I was in the cold silence of space, saw your petty foibles for what they are, saw that none of what you do, including the utter obliteration of this floating blue rock, is more than the universe can handily absorb and counter. None of that stops me from pining for quieter days, though, nor from regretting the losses I must bear, even though I wish it did."

"That's my lesson? That none of this matters?"

If there's one thing turtles are good at, and there are many things, it is contemplation. Time seems to stop as the old tortoise extends her neck, her throat bobbing in and out as she breathes and gazes over the landscape.

"When you're up there, it's crystal clear that we are all in this together. We are not turtles and people, but merely citizens of Planet Earth. It doesn't matter what gods you pray to, what clothing you prefer, what schools you go to, what languages you speak, what skin color you wear, what gender you inhabit. Turtles and roaches, crabs, spiders, protozoans, and people, we're all just Earthlings."

The tank that has been hiding behind the ridge appears, trundling across the valley with surprising speed, its turret rotating as if seeking a target. At one point, it is aimed right at us. The sight of it makes me shiver.

"You realize the irony here, right?" I ask. "I mean, having this conversation about perspective and world peace in one of the most hotly and long-contested places on Earth?"

"Irony is like pornography," the turtle immortal grunts. "Hard to define, but we know it when we see it. Glad you picked up on that sweet detail. We worked to arrange it for you."

The tank disappears from view. The rumbling fades. The sky remains free of war planes, but heavy clouds drift in.

"Soon there will be spring wildflowers, my favorite food," the old tortoise sighs.

And then the skies open up and the rain falls like crazy. Rivulets appear on the ground and the colors of the distant peaks change. Plants rise up so fast I can see their leaves unfurling. Berries gain color right in front of my eyes, and spring, as the old tortoise has predicted, arrives.

She starts to bop and wiggle and shimmy. I smile at her dance but, before I can compliment her moves, I'm back in my park.

The wind is still blowing.

DAO IS BIG; WE ARE SMALL

Although some Daoist practitioners assert that the hours between eleven at night and two in the morning are best for meditation, I find that doing my internal work at the start of my day assures that I won't put off the practice and that it won't fall pretty fast to opportunities, conflicts, and demands that life naturally and regularly brings. Despite this preference, today I find myself starting practice in late afternoon. It's not yet dark, as it is summer and the sun the sky holds onto like a stubborn child, but the light is dim.

Perhaps it's that twilight energy that lands me in jungle water up to my knees, a thick tree canopy blocking the sky, a vast river some distance away. Man-sized ferns and broadleaf foliage abound. My glasses are fogged over and I wipe them clean on my robes, only to have them immediately fog again. Mosquitos create a halo about me. A light drizzle falls. When I take a step, mud clutches at my shoe, the ground shifts, I lose my balance and slide gently down to land on my hands. Large bubbles erupt from the water and a low voice addresses me.

"Please restrain yourself, Monk. You're scaring the other monkeys."

Before I can regain my feet, I'm dragged forward by some inexplicable force. I don't fully grasp the situation until a giant head breaks water in front of me, eyes the size of coffee mugs, a snout more or less the length and girth of a Pekingese lapdog, and a neck as long as the two-handed battle sword I use for tai chi practice. Unlike most turtles, whose heads retract straight back like a preppie donning a tennis sweater, this one's neck curves asymmetrically. I'm sitting atop a giant aquatic turtle of the sub-order *Pleurodira*, the so-called side-necks. The group is represented only in the southern hemisphere these days, which means I'm likely somewhere in South America.

"What are you?" I ask, a bit intimidated by the behemoth beneath me.

The turtle lifts the rear end of its shell and exudes an impressive, mushroom-shaped sex organ.

"Wow," I say. "I wasn't asking about your gender."

He looks back at me and pulls his beak into a grin.

"We Daoists are known for our sense of humor," he says. "Immortals even more so."

"Nice," I say. "But really. I can't think of any species of turtle as big as you, and I thought I knew you all."

"I chose this body even though I am the last of my kind. My compatriots died off five million years ago in the Miocene Epoch. I'm not sure what variation of my DNA has let me live so very, very long, has let me see so many countless, tedious sunrises and sunsets, the passing of so many ages of shallow seas and frosty shores, and all the giant mammals, the megafauna that crossed these lands. You cannot imagine the *Megatherium*—"

"The giant ground sloths," I say.

"And the *Smilodon*—"

"Saber-tooth cats," I interrupt.

"—that came in the night," he finishes, shooting me his best giant sideneck turtle's version of the stink-eye. "And the *Arctotherium*—"

"Short-faced bears."

"They'd scoop a turtle's guts right out with those giant claws, do it without even rending the shell, just go in right at the soft skin of the shoulder or thigh. Of course, there were friendlier *Stegomastodons*, who turned up the soil with those elephantine tusks, loosening the earth, so our females had an easier time digging in eggs. They also kept the *Macrauchenia* away, those foul-tempered spitting ungulates—"

"Proto-camels," I say.

The turtle seems increasingly annoyed at my interjections, but in characteristically turtle fashion, he stubbornly soldiers on. "The *Doedicurus*, with those formidable, club-like tails, could splinter our shells if we made a bid for a plant they liked."

"Giant armadillo-like glyptodonts," I mutter, beginning to worry that I may not be making a friend.

"*Hippidion*," he says. "They were kind to me. They sang in the most divine way, more so than any songbird, and were wont to stir a female of my kind to rapturous appetite, all the better for one like me."

"Extinct horses," I add. "Closely related to the wild ones of today's American Southwest."

"More agreeable than the *Toxodons*," he says, his beak closed so tight, I'm amazed any sound gets out. "Those were so big and grumpy and often trampled us underfoot without even noticing."

"The bow-tooths," I say proudly. "Large members of the rhinoceros-like *Notoungulata*."

"What a horrible person you are," he hisses. "If I'd known you'd be this way, I would never have agreed to teach you. Actually, I'd like to drown you dead for the way you interrupt me. I'd like to climb over you and press you down into the mud until you run out of breath and die. Any monk as rude as you deserves that kind of treatment."

"I'm so sorry!" I say, thinking that I seem to do almost as much apologizing as learning in my spirit-writing sessions. "You're just such a marvel, I got carried away."

"No excuse at all."

"Hold on!" I shout, clapping my hands together. "I know who you are."

"Typical noisy monkey. After all your pompous peacockery, I'd be surprised if you didn't."

"You're *Stupendemys*! Largest of all freshwater turtles, larger even than the marine giant, *Archelon*."

"Who was my flighty cousin, by the way. Never could get him to sit still at the seashore long enough to talk story."

"And we're in the Amazon?"

"If we were in the Amazon, you would know. It's one spectacular river, even now. We're off in a tributary in rural Brazil. The country's going to hell all around us, which I know because of how they take the timber and argue politics and chase my people."

"Your people?"

"Forget I said that."

"Listen, I'm sorry. I didn't mean to be rude with all the Latin names. I just got excited. I loved vertebrate paleontology back when I was a college undergrad."

Suddenly, I hear a familiar buzzing sound. The last time I heard that sound so loud and close, I was ten years old and had just fired a rock at a wasp nest with a slingshot. It didn't go well for the wasps, as it was a relatively big rock and a relatively small nest, but it didn't go well for my buttocks, either, which was the part of me upon which the wasps took their revenge. They chased me across an entire meadow. I screamed the whole way, but they were neither off put nor deterred. I received at least twenty stings and couldn't sit down for a week. Later, I learned that wasps can fly at speeds up to twenty-five miles per hour, which left juvenile me feeling like a superhero for having evaded them for as long as I did.

These wasps aren't angry, though, or at least not at me. They pass overhead in an arrowhead formation, the outline of which is so precise it could only appear in a session with an immortal. I duck even though they are far above me, and eye the water just in case they change trajectory and I have to dive in.

"Bones don't tell you everything," the immortal replies. "They don't tell you how clean the air was back then, for example, nor how everything existed in such beautiful balance before you terrible monkeys ruined everything. Now the air reeks of burning trees, chain-saw oil, and tractor diesel, but back then it was fragrant with flower and leaves and fresh soil and the sex glands of females."

"You really have a one-track mind."

The immortal chortles. "Maybe so, but if you're going to have one track..."

He leaves the mud and treks overland toward the river. I follow in the broad gouge made by his dragging plastron. Soon, I realize we are not alone. The stripes and blotches I had taken for exotically colored jungle trees are actually small indigenous people with painted bodies, dressed in loincloths, their breasts bare.

"These are your people?"

"They are," says the giant immortal. "They worship me as a god."

"That's nice, but they're carrying blowguns and spears."

"Yes, and their darts are tipped with enough frog poison to dispense with many an anthropologist and logging scout. They won't bother you, though. Nobody can see you but me."

He slides into the river and the mud falls from him. Washed clean, his head is as bright yellow as a ripe banana and his shell shows red lines over a blue background of a hue I've seen only on gaudy tropical birds. He luxuriates in the current, clearly enjoying himself, his eyes closed. A six-foot caiman swims close. The giant bites off half its tail, leaving it to splash, wounded in the water. The worshippers watch from the bank, pointing. Free from the camouflage of the bush, I see they are not quite pygmies, but they're not going to be entering any basketball tournaments, either. They debate whether or not to retrieve the wounded reptile and have him for dinner, but before any decision can be made, the river erupts in fish, razor-toothed piranhas that reduce the caiman to bone.

"I feel badly for him," I say.

The giant immortal does a turtle approximation of a shrug. "We are all part of the eternal Dao."

"Thank you for reminding me."

"Sarcasm doesn't suit you, Monk. He died because I elected to take a swim. The part of him I didn't eat fed not only the piranhas, but the invertebrates living in the mud of the riverbank and the crustaceans at the river's bottom. All those creatures are part of a larger system, which has them consumed and renewed and reborn and reproducing in the steady churn of life."

"There's a saying from the Daoist Classic of Purity, *Qing Jing Jing*—" I begin.

"—*Da Dao wu xing*." he finishes. "The Dao is big."

"I guess I should have figured you'd know the phrase."

"I'm the one who came up with it," he laughs, playfully cupping his massive forelimbs to splash his followers on the bank. "And I'll tell you something to go with it. At a certain point of scale, changes in quantity become changes in quality."

His people run in and out of the water like children in an amusement park pool, giggling and laughing.

"You enjoy these people," I say.

"I enjoy these quiet Amazon backwaters," he answers. "I enjoy water hyacinths. I leave the other watercourses to my side-neck cousins, the brilliant red heads and yellow heads, and even the straight-necks. They would be diminished by my presence, and they would make too much of this one, lone holdover, and in doing so, they would lose their way."

"But the natives haven't lost theirs," I gesture at the aboriginals.

"Indeed not. I move them from jungle patch to jungle patch to make sure they stay safe. I keep them from those who would destroy their way of life without ever stopping to learn their wise ways, their balanced connection to the land, the way they can traverse levels of reality using their plant preparations, their ritual worship not only of me, but of all that is natural and pure in the world."

All at once, the jungle vibrates with a great din. It's a profoundly disturbing sound—artificial, threatening, and wrong.

"What's that?" I ask, even though I think I know.

In response, the giant turtle rises from the water and spins in the air like some bizarre alien top, his tail pointed straight down, his massive snout to the sky. It is a strange and unnatural sight, and it is clearly some kind of warning to his tribe.

They waste not one moment in responding. Mothers gather children in arms, teens melt into the forest, and men stuff poison arrows into their botanical tubes. Then the whole tribe just disappears, leaving no evidence at all that they were ever there, no branches broken, no footprints in the ground, not so much as a trodden leaf or stem.

A moment later, the loggers appear. They wear hard hats. They bellow at each other. Behind them come trucks and bulldozers with tires as tall as giraffes. Birds take flight, their cries barely audible over the roar of the equipment. Every creature that nests, crawls, slithers, or runs is crushed before the mechanical monsters. Thousand-year-old trees fall like toothpicks. The precious bodily fluids of countless spiders, lizards, sloths, snakes, and mice stain the ground.

In the face of this horror, the turtle disappears silently beneath the surface of the water. I follow him as he traverses a series of linked streams, going deeper and deeper and farther from the main course of the Amazon. I catch glimpses of his people on the banks beside us, but only because I know to look for them. At length, we come to a new glade, a place of dense foliage and fragrant aromas. The giant turtle evidently understands *feng shui*—the perfect confluence of water and wind and brightness and shade, of tall stands of forest and short undergrowth, of the way the soil veritably glows with vitality—for there is something most magical about the site he selects for his chosen people. Appreciating it, too, they rejoice in dancing and singing, copulating with delicious abandon, splashing about in clean water, all under the watchful eye of their god, who will brook no interference of their joy from a toothed reptile or fish, no unwanted intrusion from any parasite.

"So, Monk. What do you notice? What have you learned?"

"They love you. You love them."

"Beyond that."

"Terrible things are happening to the jungle."

"And why is that?"

"Material appetites?" I venture cautiously.

"Yes. The biggest mistake a human can make is to fixate so much on the material world, to cherish *things* instead of each other, instead of feelings and thoughts and ideas. The material frenzy is what brings loggers here. Look at my people. By material standards, they have nothing. And yet see how happy they are. Can any of your modern friends match their sheer joy? Can you?"

"And yet sometimes people do come together around energy, and it takes a great deal of energy, a great deal of coherence, to create a material thing," I venture.

"True. And so what? It takes a great deal of energy to produce a pile of shit, too. You have to find food, chew it, swallow it, pass it through you, and expel it. Does this make it somehow special? What's special is the life it represents. Too much focusing on the feces, you miss the pleasure of the meal. Remember, attachment is merely the fight against change, when change itself is the only true constant. Embrace it!"

"It's funny to hear you talk about impermanence when you are the very epitome of endurance," I say.

"Duration does not obviate impermanence. In fact, it is precisely because I have endured so long that I have come to see the true nature of everything."

I sit down in the water and gather my knees to my chest. I understand what he's telling me, but I have trouble coming around to actually living these ideas. If I can't live them myself, I wonder how I can convincingly spirit-write them.

"How can I live this way?" I ask.

"Let's try this. Can you imagine walking with a stick? Do monks like you walk with sticks?"

"Some monks have staffs," I say.

"Good. Now, imagine that every step you take, you become smaller and smaller against the backdrop of nature. Eventually, you are reduced to a tiny creature. Your stick changes, too, becoming ever smaller until it is a mere toothpick. You keep on going until you are reduced to the atomic level. At this level of scale, a cotton-swab would appear to you to be a kilometer long. Here, a snipped-off hanging fingernail could crush an army. Here, the true immensity of the cosmos is revealed."

I hesitate, still feeling that this may be one turtle whose thinking is too deep for me, whose message I cannot fully comprehend. "How does this help?" I ask.

"Must I spell it out for you?"

"I'm afraid so," I say weakly.

"If your lives are so small and insignificant, aren't your impulses and urges equally so? Do you think your desire for a pretty sports car warrants the sort of devastation you just witnessed? Your need for a second story on your home? Your desire for a wooden stand on which to place your big-screen television or a hardwood desk for your computer, the very devices that encourage you to escape this world? Do you not see the irony in ignoring and destroying the very rewarding paradise for which you search and yearn?"

"I do," I say.

I leave the immediate company of the ancient giant then, and float betwixt and between his followers. Close to their faces, it is clear that, despite their nose rings and their earrings and their bangles and loincloths, face paint, bare feet, prehensile toes, narrow shoulders, taut bellies, and oval eyes, they are not so very different from most other people I know. I see in them the same need to believe in something bigger than themselves, the same desire to nurture their children, the same wish to preserve their culture and way of life. I see in them a reverence for nature familiar to me from Daoist teachings, but also a refreshing absence of the desperate grasping that so often plagues seekers in our speed-and-greed Western world.

I notice, with a shifted perspective, that even the oldest members of the turtle tribe move smoothly, deliberately, and without apparent encumbrance or effort. These elderly people are at the core of the community, nurtured and supported by all the rest, laughing and dancing as much as anyone. There are other signs of enlightenment that I took at first to be of a primitive society, but now realize aren't, namely a lack of material excess, the ring of freedom, the joy of sharing, the satisfaction of resting, the pleasure of artwork already appearing in the form of paintings on bark made from flower dyes, ceramic dishes being fired in pits, and cauldrons already dug and built, even though they have only just arrived at this, their new home.

What I don't see is suffering, even among the injured or ill. Sensing their keen sense of balance, wildlife does not avoid them. Already, despite the short time at this camp, birds swoop by and are offered seeds from fingers, rodents run past but are petted not shunned, tarantulas walk across sleeping hammocks and are gently returned to the ground, serpents, undisturbed, coil in baskets and sun themselves in bright patches of ground.

The ancient one eyes me for just a moment, then returns his full attention to his people. He floats, weightless on the surface of the stream he has chosen, as little fish clean the algae from his shell and crabs trim his claws. He watches his followers with attention, devotion, and love. I find myself wishing I could be one of them, but know it is time for me to go.

In the distance, I can just make out the brown wisps of exhaust smoke rising from the earth movers. If I pay close attention, I can see the direction in which they are moving.

They will be here soon. I, however, will not, as before I know it, I am back in my park. Darkness has fallen.

SHAMANISM, SCIENCE, DREAMTIME, AND DAO

My hometown is under a hurricane watch today. The weather is not yet in any way visibly threatening; in fact, the calm before the storm is peaceful and quiet, but the air is heavy with water, sea salt, and panic. There are three-hour lines at all gas stations—customers are filling not only their cars, but cans so as to be able run home generators— and shops are out of batteries, water, peanut butter, crackers, soda, milk, bread, and handgun ammunition. I am alone in the park. In the distance, I can hear car horns and shouting, but closer to me, it is eerily quiet because all the birds, knowing better and sooner than we do what's coming, have flown away. Curiously, I see several snakes cross the bicycle path. They move in leisurely fashion, even taking a moment to soak in the heat of the asphalt, apparently knowing they may not see the sun for a while and knowing, too, that there are no hawks, shrikes, falcons, or eagles about to swoop down and eat them right now.

I feel giddy. It's hard to explain, but to meditate in the face of a storm, to prepare my mind for chaos, is a very satisfying thing to do. I'm not one to leave things for the last moment, so I've already done my shopping and boarded up my place and already lent a hand to my students and followers and friends, as well. The time for helping others has temporarily passed, though it will come again in the aftermath of the storm, the rescues, the cleanup, the rebuilding. Right now, though, as I stand in the shade wondering if this favorite tree of mine will survive the upcoming gale, I can see the natural energy of cycles, calm and storm and calm again, and can watch what my mind does in its little fits of imagination and anticipation.

Perhaps because I've gotten used to the low barometric pressure in the park, I don't notice that I've meditated my way into another pending storm, though this one is far from home. In fact, in my trance, I float over a busy coastal city redolent with the cuisines of India, China, Japan, Thailand, Malaysia, and Indonesia. It's a heavenly mix, slightly salted by the sea air. There's a stiff wind blowing and ominous clouds over the sea. The storm is much closer here than it is back at the park, and I watch with fascination as the tropical tide recedes, leaving a vast plain of wet sand. As I drift lower, I catch sight of a newsstand. Headlines declare a killer typhoon is looming and that a Canadian woman has just been "taken" by a saltwater crocodile on the town beach. I can tell by the fact that the very few people out in cars are driving on the "wrong" side of the road that I'm in Darwin, Australia, on the Indian Ocean.

I drop like a drone until I am just above street level. I fly over the tourist zone and small high-rises in the center of town, then a few blocks in from the waterfront. Patrolling police cars blare storm warnings through car-top speakers, but things are much more civilized here. The queues at stores are patient and quiet; nobody is shouting. I see good Samaritans conversing quietly with homeless people and pointing out shelters in local postal offices and churches. Perhaps because the town is much smaller than my own and there are more than enough fueling stations, the lines for gasoline are moving more quickly and are not nearly so long. I try to envision everything I see being washed away, but I can't quite conjure the image. All seems so stable, even ancient, colored, perhaps, by my decades-old fascination an aboriginal culture at least 50,000 years old, so very much more ancient than either the Abrahamic or Daoist traditions.

I am drawn to a two-story corner building. Below, there is a seafood restaurant with foreign and domestic beers displayed in the window. Above, there is an apartment with a balcony overlooking a wide street. The balcony features a lush array of tropical plants, a wooden chaise lounge, and a rose-breasted cockatoo in a cage. A gray plastic bussing tub sits on a two-top dining table. Inside the tray, resting in perhaps eight centimeters of water, is a young female Fly River turtle, so large she cannot possibly swim in the box, so she keeps bumping her uniquely-shaped snorkel nose against the plastic, with a crusty scab to show for her trouble.

This species, often called the pig-nosed turtle, resides right here in Australia's Northern Territory and in New Guinea. When I interned at New York City's Bronx Zoo after high school, there was an adult pig-nose in the collection. I took care of it sometimes and loved to watch it swim, as these turtles use their forelimbs as a sea turtle does. Specifically, they pull downward with both limbs simultaneously while tucking the rear legs out of the way to zoom through the water faster than other freshwater types.

As I grow closer, the cockatoo tilts its head and glares at me.

"You can see me?" I ask, somewhat stupidly.

By way of answer, it issues an earsplitting shriek.

I get closer to the turtle. She looks up at me.

"Hello, Monk," she tells me in a sultry voice. "Interesting timing."

"I'm not aware of having picked it."

"Then we have to work on your awareness."

"Fine," I say. "Great. I just meant I don't exactly choose these spirit-writing sessions."

"Of course you do."

"Right," I sigh. "Of course I do. So tell me, what are you doing in that tub? I gather you were captured?"

"I suppose that's the word. I was only a hatchling and naïve to the technique of using beef to bait a hook. All I have now are hazy memories."

"You can't even swim a lap, can you?"

"A lap?"

"One end of the plastic tub to the other."

She shakes her graceful head sadly. "I could when I was first brought here. Three paddles one way, turn around, three paddles the other way. Now the moment I take a stroke, I'm done."

"I'm so sorry," I say.

I hear a siren wailing and turn my attention momentarily to the street. An ambulance goes by and stops in front of a townhouse less than a block away. Paramedics emerge, offload a gurney, and wheel it to the door. I watch as they go in and come out again, this time with an elderly woman in their care. She looks dreadfully pale, sunken into the sheet as she is. I know that back home, people often suffer heart attacks in anticipation of big storms, and sometimes at the height of them, too, especially if there are tornadoes. I wonder how many other quiet, invisible tragedies are unfolding in town right now.

"There are two sides to everything," she says. "Yin and yang, right? If I hadn't been captured, there is much I would not have learned. My life would have been narrower and poorer, even though I miss the sweet water of the river."

"At least you're safe from being eaten. I read somewhere that the aboriginal people especially enjoy the fatty skin around your neck."

"They call us *Warradjan* and my sweet-water sister, *Nadwerrwo*," she answers. "We stayed in the rivers when the seas retreated from our land, unlike my other sister, *Manbirri*, who paddles as I do, but stayed at sea."

"*Nadwerrwo* is the Northern snapping turtle," I say. "*Elseya dentata* in the language of science."

"Scientists are interesting cultists. Do you think they believe in souls?"

"I bet some of them do but won't admit it."

"Well, no matter how many legs we have and no matter the type of skin, feathers, skeleton, or scales, we are all gloves enlivened by the same cosmic fingers. When we die, it is not the glove that counts, as it is discarded and allowed to return to the Earth to provide material nourishment to the next glove, but rather the hand."

"Gonna rain, gonna pour, gonna rain, gonna pour," says the cockatoo. "Gonna drown, gonna scream, gonna die."

"She's such a drama queen," the turtle remarks drily.

"Well, a typhoon is on the way."

Just as I say that, portentous raindrops begin to fall. Soon, they are pounding the tin roof above us.

"We don't have a lot of time here," I say. "Is there some lesson you want to reveal to me right away?"

"Tell me your Daoist lineage first," says the immortal turtle.

"You don't know it?"

"You think I keep track of every detail of every spirit-writer I help?"

"Of course not. Presumptuous of me. I was ordained in the *Long Men Pai* line of the *Quanzhen* sect—"

"Very open and inclusive," the immortal interrupts.

"Yes, but I mostly follow the *Shangqing* branch. Some people call it the *Mao-Shan* path of Great Purity. It's a blend of Northern and Southern Daoist camps from the fourth century of the common era, although parts of our tradition can be extended well back to before our calendar year 0. The tradition is based on a message given to an enlightened woman by immortal teachers, then propagated by her children and grandchildren."

"So you've got spirit-writing in your direct line."

"I do."

"Well, the inclusiveness is good because my message is about that."

"A link to Australian aboriginal beliefs?"

"How did you guess?"

"Well, we *are* Down Under after all."

"Let me out," says the cockatoo. "You two are going to die talking."

"She won't give us a moment's peace until you do it," sighs the immortal. "The key is on a white plastic hook on the wall just inside the sliding door."

I go inside the apartment even though the door is shut tight. The place would be neat as a pin were it not for the books. Titles on philosophy, anthropology, evolution, sociology, aboriginal culture, mythology, and religion spill off the bookcases, which cover every inch of available wall space, into stacks on the floor. The contents of the refrigerator tell me the missing professor is a vegetarian with a keen preference for South Indian cuisine—frozen dosas and the like. I take the key from the hook on the wall and go back outside, where I find the cockatoo hyperventilating in fear, her little black club-shaped tongue protruding from her open beak. I open the lock and then the door.

The cockatoo steps out onto my hand. Just then, a blustering gust knocks the cage down and away, spilling birdseed everywhere. Three top-heavy potted plants fall like dominoes. The cockatoo shrieks in fear. I can actually feel the barometric pressure falling.

"Not yet," the immortal warns. "Just wait a moment and the rain will stop. Storms come in bands."

We remain there together in the company of the wind, the immortal placidly regarding me from her bussing tub, the cockatoo grabbing my fingers with her claws as if trying to crush them.

"My captor is a university professor," the immortal says conversationally, "a philosophy teacher specializing in comparative religions."

"Can't be much of a thinker if he left you and the parrot here to drown."

"He's working in the bush at this time, researching the Dreamtime of the aboriginal people. His girlfriend is supposed to come in and look after us, but I guess she panicked or forgot."

"What is the Dreamtime, exactly?"

"It means the uncreated period before all things," says the immortal. "The Daoist analogue is what you call *wuji*, the time before there was a universe, when there was only empty space pregnant with infinite possibility not yet realized or born."

"I met the immortal who coined that reference to pregnancy," I say.

117

"I know you did."

"So this is a lesson about the universality of shamanic beliefs."

"For God's sake!" screams the cockatoo. "Take your thumb off my leg and let me fly. If you two want to get swept aloft in a tornado or drowned by a tsunami, that's your business. I've got to get out of here!"

Indeed, the wind has died down and the rain has all but stopped. I crane my neck to look beyond the edge of the building and see that the sky off toward the beach is as black as night. The streets below are completely empty now; not even a patrol car prowls. I thrust my arm out into the light, light drizzle and take my thumb off the bird's foot. She gives the turtle a quick look, nods goodbye, and takes off to the south, away from the ocean. I see her buffeted about by the winds up high until she is lost from view.

"When you use labels like aboriginal or shamanistic, we forget that we're just talking about the observation of nature and natural phenomena," says the immortal. "At its best, science is no different."

"They call that basic research," I say. "Most science these days is driven by commercial interests, but there are still some people making essential inquiries into the nature of what is."

"They're the ones who will most appreciate your spirit-writing," enthuses the immortal.

"Uh, I don't want to contradict you, but I think they have their own ways of doing things. They don't talk to turtle spirits much."

"Doesn't matter. If they're truly open-minded, they will take the information where they can get it."

"I suppose that's right," I say, agreeing in theory, but dubious about the gritty challenge of mixing spirituality and science.

"Any random citizen of Plato's Athens would be more than a match for a corresponding modern New Yorker," declares the immortal. "In those days, people weren't quite so blindered. They made their own observations about the natural world, based on their own curiosity. Nowadays, everyone just takes what they're told for granted."

"Buying into the narrative," I say.

"Just so."

The wind stops completely. I lean over the rail again and can hear the Indian Ocean crashing angrily ashore.

"Won't be long now," I say.

"The interesting thing is that there are multiple narratives," says the pig-nose immortal, "and not just amongst humans. Whales have their own, singing it across thousands of miles of ocean. We immortals have ours, too, and it spreads across our realm. Trees have one, connected as they are by their roots underground and by the mycelium that receives sugar as payment for transmitting information from tree to tree. Insects share their view of the world through chemical signals."

I've already been told about the importance of story, but this idea of species-specific stories and worldview is different, as is the idea that so many human narratives are just the same story told by different tribes and in different languages. I'm still pondering the profundity of this view when I see a tornado in the street. It swirls left and right to whisk aloft dust and debris like some celestial vacuum cleaner. Raindrops pour off the roof and create an insulating curtain on the balcony. Termites take their last bites of wood and frenzied pantry roaches eat their last.

"The storm surge is going to be as high as this balcony," I say.

"Oh, most certainly. And right on time."

"You'll drown."

"Don't be silly. I'm a turtle."

"It will tumble you. You have never been swimming in anything like this. And even if you had, your muscles are atrophied from captivity. Let me rescue you. I flew here, or at least I sort of floated, so I should be able to float away with you."

"You appeared," says the pig-nose immortal. "And you're going to disappear in a minute or two. I don't need rescuing, but I would appreciate it if you would take me to the loo."

"The loo? Don't you just, um, relieve yourself in the tub?"

"This isn't about that, Monk. The loo I mean is the innermost room in the house. The walls will take the brunt of the storm. As the water recedes, I will follow it. I will let it carry me to freedom, and thence to the street, where I will swim south to the river and from there to my home down Kakadu Way. Hurry now. There isn't much time."

I pick her up. She is full of life, hard-muscled, dense, and strong. I touch her nose gently and the scab falls away, revealing bright, pink skin beneath. I take her to the bathtub and lay her gently down. She looks uncomfortable on the dry, cold porcelain.

"Do you want me to run some water?" I ask.

"No need," she says. "Water aplenty will be here soon enough. Please shut the door behind you when you leave."

"Remember that water is unpredictable," I say. "It never has a plan."

"And neither should we. Goals and strategies, yes; plans, no. Life is too changeable to rely upon them."

I bid her good luck and return to the balcony. I sit down in the professor's chaise, which I now recognize for a pearlers' chair, the sort divers use to rest on the decks of boats after they have plunged into the sea seeking shining orbs to feed their families, always hoping for that fist-sized treasure that will change their fortune. I feel I have found my own treasure in this Dreamtime with these turtles. I sigh as the storm surge approaches—a veritable wall of water. I look it up and down and frame it with my fingers as if to capture it like a photograph. When it is too close to fit inside my grasp any longer, I float up off the balcony and stand upon the roof.

The waves subdue the sky. The wind hits me hard and turns me white with sea salt. I float a bit higher. The roof flies off the apartment and the waves wash away the restaurant, the whole building disappearing into the sea. The surge remains in place so long, I fear the pig-nose will not survive it. I hold my position, unwilling and unable to look away from the apartment.

At long last, the surge lowers. As it does, I see a small black shape riding the whitecaps like a surfer, rhythmically pumping her legs, flying through the foamy debris. She navigates the chaos, even though there is so little left that is solid, so little left of what was before. She clearly knows how to work the current, how to proceed with seemingly effortless efficiencies. It appears as if she covers a mile with every stroke, heading south, right back to where she wants to be.

I want to congratulate her, to tell her how very happy I am that she has made it out alive, but I begin to evanescence before I am able to do so.

It doesn't really matter, because she already knows.

She knows *everything*.

And I am back in my park just in time for the storm.

NATURE TEACHES THE LESSONS WE NEED

One thing that many people don't understand about meditation is that there is no such thing as putting a stop to thinking. If we stop thinking, we die. All we can do is either watch our thoughts or direct them; ending them is not an option. All the same, we can learn to focus better, to concentrate on one thing at a time, and to consider feelings and ideas more deeply, with less judgment, and with greater patience and forbearance. To hone the mind this way, to counter the habit of jumping rapidly from impulse to impulse, thought to thought, feeling to feeling, is to counter the effects of using our digital devices, for these devices depend upon and encourage precisely such saltatory habits.

If we are successful in quieting our minds this way, we can find contentment and peace. Despite the romantic ideal of meditating steadily as the world falls apart around us, as bullets fly, bombs fall, and fires rage, achieving mental equilibrium is easier when it's quiet outside. That's one of the reasons I go to the park. Most of the time it is quiet there, I don't bring a computer or phone, and the natural energies of the trees and air and water help me to relax as well. That's why there is a great Daoist tradition of meditating in caves, on mountaintops, in or near waterfalls, or by lakes, ponds, and streams.

Today, however, my meditation is disturbed by a terrible dogfight. I've had the experience of losing a little dog to a big one—a Chinese Crested that was eaten, quite literally, by a neighbor's Shar Pei—so

I know how quickly most such encounters end. This one, however, lasts and lasts. Two pit bulls are involved, dogs that have been bred to fight and to endure grievous injury. The altercation breaks out not a stone's throw from my favorite tree, and the owners, both aggressive young men with tattooed skin and wearing heavy chains around their necks, seem more concerned with shouting insults at each other than they do with pulling their dogs apart. The ground is soon dotted with blood. The growls and whines are both frightening and pitiable, and are themselves drowned out by the angry shouts of outraged onlookers insistent upon an end to the contest.

It does end eventually, with both hounds limping off, but it leaves me shaken, my pulse high, my breathing rapid, my mind awhirl. Perhaps that's why it takes me a good half hour to get into the frame of mind for an encounter that will end in spirit-writing. When I am finally thusly entranced, I find myself sitting beneath a graceful acacia tree in the shadow of Tanzania's Mount Kilimanjaro. Elephants stroll by. I know they're fearsome beasts, but I don't fear them. I have, after all, stood up to a typhoon. Nearby, a lion gnaws on a giraffe carcass. Laughing jackals circle. Hyenas gather in overwhelming numbers. A lioness emerges from the grass, sails over the hyenas in a magnificent leap, and joins her mate. Vultures appear in the air. A tawny eagle tumbles down, huge wings hammering. The moment he lands, the hyenas open their circle to provide the lions an exit, and the big cats oblige.

A twelve-kilo leopard tortoise wanders up, its eyes bright, its domed shell a patchwork of black and yellow. He regards me thoughtfully and squeezes out a dry turd.

"We don't call the eagle 'the emperor of the air' for nothing," he opines by way of introducing himself. "He sees everything from the great heights that are his home. Tell me, Monk. When did you last fight for a meal?"

123

"I'm not sure I've ever done so."

"Really? Well, no one that lucky should indulge a mood so blue."

I stand up and stretch, feeling heavy and stiff. "I'm worried about the world," I say, "and my visits with your friends aren't easing the feeling."

"If Kilimanjaro doesn't cure you of that, nothing will," he says, his voice a low rumble. "Did you know that the west slope is called the House of God?"

"For good reason," I allow. "It's a majestic sight."

"Indeed. Whenever a flood threatens to submerge me or I lose a mate to a stronger tortoise, I gaze at Kilimanjaro for solace."

"And yet all it does for me is make me cry that such beauty will fall to my kind and our appetites."

"Acceptance should be a monk's trait, no?"

"It's not my strong suit. I'm more likely to try and fix things than I am to let them be. I cling to the hope we will someday soon change the way we see each other and the world, and as a result of that new sight, change what we're doing to the oceans, mountains, deserts, jungles, forests, rivers, and streams."

"I noticed you didn't mention the worlds underfoot."

"I'm not quite so concerned about caves," I say. "They seem pretty safe from our interference."

"Human poisons trickle down," he says. "Let me show you something."

We leave the shade of the tree. The sun is hot in a way I've not felt before, not the glistening, muggy blanket I find in Southeast Asia and not the distant touch of a North American spring, but a blazing knife that finds my vulnerable places—the back of my neck below my square monk's hat, the tops of my hands, my forehead, and what my tactful Chinese friends call my "imposing and characterful" nose. After a few minutes, we come to a structure that reminds me of a sandcastle once assaulted by the sea. The tortoise stops at the edge of the thing and gazes at it. It is taller than head-high and as wide at its base as a dining room table for twelve, but the lack of sharp or defining edges are straight out of an Impressionist painting. The flat top would make it look like a tiny dormant volcano but for the fact a stream of bugs is entering and exiting a frisbee-sized hole there. Careful not to squish it, I pick one up. It is brownish in color but with an orange hue. The head is nearly half the size of its striated body and is equipped with both antennae and two sets of pincers, one set above the other. It writhes in my hand, trying to get those pincers into me.

"You're holding a mound-building termite," the tortoise tells me. "The above-ground mound you see here is its nest. Most of it is underground and connected to other mounds in what you might call a termite nation. There's actually a species of South American termite whose nests are even larger, as big as the pyramids in Egypt. Imagine a human city turned upside down so that rooftops point toward the Earth's core. Working cooperatively within caste systems that make them incredibly efficient, those little bugs excavate galleries for breeding and food storage and orient connecting shafts in accordance with prevailing winds, magnetic fields, and the angle of the sun so as to keep the whole mound comfortably cool. These nests are thousands of years old, though climate change threatens their continued existence."

"How do you know all this?" I ask incredulously.

"You ask this of an immortal?"

"Right. Sorry."

NATURE TEACHES THE LESSONS WE NEED

"Think of it this way. All the termites are ferocious and single-minded in their application of their energies to projects and principles that are required for the common good. They speak as one because they think and act as one. As I said, there are worlds within worlds within worlds on this planet—millions of species unknown to you, deep in the ocean, too—and humankind thinks so rarely of them, if they think of them at all."

"I've continued to hope we can manifest some kind of spontaneous evolution for the benefit of all sentient beings," I say.

The tortoise turns his back on the mound and stares up at the snows of Kilimanjaro, the loose skin of his throat pulsing in and out. I follow suit. I try to imagine the great mountain turned upside down, with people scaling its summit by digging with spades.

"We need to take a little trip," he says at last. "Climb aboard my shell so we can travel together."

"I'm awfully heavy," I say, looking down at him dubiously.

"Again you doubt me."

I summit his shell with lightness and precision, grabbing the margins of his shell to settle myself. He shifts a bit to set his balance, then begins to walk. At first his strides are small and typically turtle-like, but soon they lengthen and speed up. Before long, my perspective on the world changes. Trees shrink, zebras assume toy-like proportions, and water buffalo shrink to the size of Matchbox cars. We're still on the ground, but not attached to it. Clouds seductively caress the naked flanks of Kilimanjaro. Invisible beetles chatter and tiny warthogs grunt underfoot. I sniff the musk of big cats, the pungency of elephant dung, and a minty note on the wind coming off the mountain. We zoom over the plain. The horizon exerts a tide-like pull.

"Tell me where we're going," I say. "If I'm going to keep this up, I have to know."

"Well then, we're going to Olduvai Gorge."

As if on cue, the famous crater appears. We stop at the raised and craggy perimeter, which is marked by drier soil than that of the tree-filled plain where we started. The landscape gradually reassumes normal proportions.

"My friend, the famous anthropologist Louis Leakey, spent his life unearthing the secrets of this place," the tortoise rumbles. "Let me share what he showed me."

Without warning, he leaps off the edge of the cliff. We slide down, taking more than a few bumps and scrapes along the way. Finally, we come to rest in the scrub at the hard bottom.

"Millions of years ago, this was a bog," the tortoise says. "Leakey called it the Oldowan Slaughter-House. Thousands of creatures lost their lives to its hungry mud."

I toe the ground with my shoe, knowing that below me lies the fossil evidence of the progression from one of my forebears to another, monkeys, apes, hominids. Here lies the tale of discovery of fire and the first tools and the earliest language, the dawn of community, and the earliest impact of human culture upon the rest of the natural world. I wonder what my Daoist masters would say about this primordial landscape. I wonder how they would feel standing where I stand, knowing that migrants from here would, at the end of an impossibly long, arduous, and mountain-crossing migration, eventually become shamans in the land that would be China and, thousands of years after that, become the first Daoists. The magnitude of time's long run leaves me feeling insignificant, tiny, ignorant, and more than a little humble.

🐢 NATURE TEACHES THE LESSONS WE NEED 🐢

My reverie is interrupted by a low growl followed by a gasp from the tortoise.

He has been self-assured and professorial until this moment, but now seems a frightened child. I look around slowly, hunting for the source of the sound. I don't see it right away, for I have eyes born in the city and new to the hues, textures, and topography of this gorge. I suppose I should have figured out, in advance of actually catching sight of the leopard, that nothing would frighten my companion so much as the sight of his namesake. There is an existential quality to our own name, no matter how common, that never fails to catch our attention.

The big cat slinks toward us, his belly barely clearing the ancient earth. The fur on the back of his neck is up, his gaze is relentless, and his paws are bigger than my hands. His tail stands erect, a black tuft at the end, and even though he seems to be moving slowly, he is making short work of the distance between us. It would be both understandable and forgivable if I froze at that moment, regardless of whether any physical harm could come to me. I don't. Instead, I bend down, pick up a rock, and throw it at him, hard. Even though I could never be accused of having any kind of a pitching arm, a rock is a rock and cat bones are cat bones, and when they meet at speed, things go poorly for the bones. The leopard seems genuinely shocked as he rises up with a screech. Perhaps he cannot see me, for he does look around desperately. I pitch a second rock, which hits him in the chest. I hear the thump. I hear the tortoise laugh, too, as the leopard turns tail and disappears.

"You may not be the brightest flower in the garden, Monk, but I'm happy for your bravery and of aim."

"He would eat you?"

"He would take me up a tree and wedge me between hard branches and let me bake in the sun until I peeked out for a deep breath. Then he would take my head off with a single swipe and use his claws to scoop out my innards from lungs to guts. I've seen it happen more than once. There is no worse way for one such as me to die, unless it is from one of those worms that devours us from the inside out, little by little, growing and growing within us even though we cannot ever see them or have the tiniest prayer of being delivered from the fate they bring."

"But you're immortal. You can't die."

"In the largest sense that's true, but we're in a certain theater here and it has rules and consequences."

I find I'm sweating. A bit dizzy, I sit down on the hard, brown dirt. The tortoise wipes a fly from the corner of his eye.

"Let me ask you a question," he says. "Which do you think was the most important advance for your kind—walking upright, growing an opposable thumb, developing language, or choosing agriculture over hunting and gathering?"

"They were all important, but if I have to choose the most uniquely human one, I must choose language."

"And would you agree that evolution is a product of language?"

"I would say the reverse."

"I know, but you would be incorrect. Evolution is a word, and a word is a unit of language."

"A specious point," I say, a bit irritably.

The immortal wanders up to the now dry and dusty death bowl and tests the ground with a forelimb. It's almost as if he's worried he might slip and become a fossil himself. Satisfied the ground is no longer soft and that it will support his weight, he ventures out onto what was once

NATURE TEACHES THE LESSONS WE NEED

the fatal pit. I follow him until he comes to rest atop a pile of bones. I crouch down as I imagine what anthropologists and archeologists and paleontologists must do, and run my fingers through the dirt, coming up with a vertebra here, a tooth there, a rib, an eye socket, the top of a femur.

"Even though we generally use the word evolution to describe the interface between biologically alive organisms and the inanimate world they inhabit, it can also be ascribed to ideas."

"Sure," I say. "Ideas are pretty much constantly evolving."

"And more quickly than physical forms, yes?"

"I suppose so, unless you're talking about antibiotic resistance in a bacterial colony or some wing trait in fruit flies."

"And do you grant that there is a gritty and real connection between the evolution of ideas and changes in the material world?"

"I think that much is clear. We come up with the idea of cars and then millions die in them as the air turns brown from their exhaust. We come up with the idea of smashing the atom, and next we create a mushroom-shaped firestorm that obliterates whole cities for all but the cockroaches."

"Would you then allow that the evolution of ideas is a form of human evolution as real as the bones in the ground beneath us?"

"I suppose so."

"And are all human ideas destructive?"

My fingers touch something under the soil. I gently clear the dirt away until I see a round, ivory-colored piece of bone. I trace its perimeter with my fingers, imagining that I am a modern-day Leakey with his picks and brushes. I uncover more and more of the bone, working it gently out of the ground until I reveal it to be a proto-human skull with a heavy brow, a forward snout, and deep-set eyes. I cradle it in my hands and marvel at the virgin territory I'm exploring. I wonder, too, whether his land will also soon suffer the pernicious effects of human activity.

"Not all," I say. "But the Western religious notion that the Earth and all its inhabitants are here to serve humans has proven to be a stupendous disaster."

The immortal nods. "The assemblage of worms and bacteria, viruses and fungi that meld with your own human cells to create each human individual is a system constantly in flux, not only within itself but as part of the sky, the moon, and the stars. Denying this has led you to an evolutionary dead end."

"I still hope there's a way out," I say. "The problem is that most people are too worried about their job, their mortgage, fixing their car, and feeding their kids to care about such abstractions."

"The way the climate is changing, none of this will be abstract for much longer," the immortal declares.

"Nature giving us what we need just the way you Turtle Immortals give me what *I* need."

"Just so."

I want to thank him for this, for the African tour, and for helping me frame the problem and the solution so clearly, but I suddenly find myself back in my park.

The pavement where the dogs fought still bears the stain of blood.

WHAT WE WANT MOST IS FREEDOM FROM SUFFERING

A major challenge for meditators is the discomfort that can arise inside the body in the absence of distractions for the mind. Such discomfort may include back pain, an urgent need to use the toilet, tension in the neck and shoulders, and, in the case of my brand of standing meditation, tired legs. Since meditation is an ongoing process and since we are organic beings not robots, the arising and falling away of such discomforts make every session unique.

This morning's experience for me is quite fraught. While the park is quiet, the weather gentle, and the air clean and clear, last night's dinner has not agreed with me and I am plagued by digestive woes. In response, one part of my Daoist training says not to push things, not to go against the body's messages, and not to insist upon practice when the body says no. Another part of me values discipline, consistency, and routine and argues that there is always something that can keep us from practice, whether it's a chore that needs doing, the demands of a boss or family member, or some departure from physical well-being.

This wrestling match keeps my turtle mentors at bay for a good half hour, during which time a pelican makes a deposit upon my freshly shaven bald head. It is an impossibly large, gooey, and noxious glob of dung, and it reeks of fish. A few passersby giggle at the sight of me, drenched, distracted, and wiping myself. The stench persists even after I've cleaned up. I can't help but think it's a heck of a coincidence that both the bird and I are commanded by what we've eaten. Perhaps it's a message from on high, perhaps a version of avian humor, or perhaps just a vote of sympathy. Either way, when I finally do enter the immortal world, I discover I am standing at a urinal inside an airport men's room. I temper my surprise by reminding myself that it's a monk's job to remain humble and accept lessons when and where they come.

The bathroom is small and reasonably clean. Leaning on her mop, a nearby female attendant stirs a bucket full of cleaning solution with a toilet brush. Her bucket says Koh Samui Terminal, which tells me I am in Thailand, on what used to be an island nirvana but is now a crowded travel destination overrun by motorbikes and tourists. Balanced upon the pitted chrome base of the flushing fixture is a pyramid of empty beer bottles. Straight ahead of me is a large aquarium set into the wall and against a window on the far side. Through the window, rippling in the filtered water of the tank and dimmed by the less-than-clean glass, I can make out a pond and a garden of thick tropical foliage and flowers. Plastic plants, thick-leaved and tied to the lip of the tank with wire, wave in the stream of water issuing from a filter box.

At first, I take the football-sized object on the bottom of the tank for a rock. I pay it no particular attention, believing that the inevitable turtle encounter must be awaiting me in the terminal outside. I finish my business and am about to turn away from the urinal when the rock extends legs and an orange head pushes off the bottom of the aquarium and hangs onto the edge to look me squarely in the eye. I recognize him as a male giant Asian pond turtle.

"Greetings, Monk," he says.

"Oh. This is a surprise. I didn't notice you in there."

"That's because I shouldn't be in here. I'm a terrestrial species. Oh, the situations we immortals put ourselves in so as to teach you what you need to learn."

"I'm sorry," I say.

"You know you're constantly apologizing, right? None of the old Daoists did that."

"Times have changed," I say. "Even the most luminous of those old masters could not have foreseen how bad things would get. The truth is, I *am* sorry for the plights I find you immortals in."

"Yet you understand it's only proper we should exploit the situation so as to get to the heart of things."

"I do. Do you eat well at least?"

"Not at all. I'm so malnourished I'd go for a lizard, a mouse, even a rat."

"A rat!"

"I've got strong jaws and I need protein. All I get are fish pellets. When I was a baby, I would hide in the leaf litter by the edge of the stream so I could stay wet and swallow worms more easily in the shallows. The larger I get, though, the more fruit I need in my diet, melons and berries in particular. Say, are you going to flush?"

Embarrassed, I do so. Then I wash my hands. Using the handle of her mop to push along her rolling bucket, the attendant walks past and out, presumably on her way to the ladies' room. The door to the bathroom clangs behind her.

"A hunter scooped me up when I was just the size of a big berry myself. Sold me to a Daoist temple as a symbol of Mother Earth and pure feminine energy," the turtle tells me.

"I didn't know there were many Daoist temples in Southeast Asia."

"Who said anything about many? It only takes one. I was in one for a while and people threw junk food at us—mostly strawberry cupcakes. Sometimes blueberry muffins. Of course, the Buddhists like us, too. I was in a Buddhist temple for a while. It's all about karma. Temple-goers pay money, buy us, and set us free. As soon as we were, though, hunters came after us. Some of my kind tried to beat the system by simply not returning home. Instead, they traversed busy highways, crossed lakes, and navigated jungle understories. In the end, however, the relentless hunters captured them anyway. No matter what we do, we end up in captivity.

"The Buddhists I know are good people."

"I'm sure real, awake Buddhists are, but those temple-goers knew full well they were being deceived. They just wanted to feel better for a little while. They wouldn't face the fact that the never-ending cycle of capture and release is just another version of the wheel of samsara they face themselves, coming back as cockroaches or worms or dogs. They willfully supported cruelty and thereby greased the wheels of torture."

"I'm vegan," I say, as if that is somehow relevant.

"The attendant gets to my tank through an access panel in that room of hers. Would you get in that room and come and free me?"

I go to the door to the attendant's room. I jiggle the handle but find it locked. It's robust and I don't see any way to force it open.

"You might have to kick it down," the immortal advises.

I confess I haven't ever done that.

"What kind of man makes such an admission?" he chides.

"You're a piece of work. My grandmother could learn a thing or two about guilt from you, and in case you don't realize it, that's high praise."

"Bring me a piece of papaya when you come—that's the taste of freedom."

"If I manage to set you free, won't the hunters just bring you back?"

"Depends where you leave me. Then again, maybe it doesn't. Maybe everything is predetermined. Pre-ordained. Fated."

"I don't think so," I say, making a practice kick at the door.

"I can't place faith in a man with such weak legs."

"You sure know how to make friends. This is an airport. Security is serious business here. This door is industrial grade."

"And here I was thinking monks were masters of martial arts."

I kick open the door. I wait. Nobody shows up.

"I could be vegan," says the immortal. "Especially when it comes to crayfish. I used to eat them, but now realize they feel things more keenly than the rest of us, despite their simple brains. Maybe it's because they hunt tiny prey in murky silt and therefore must be sensitive to every little thing in the world. They scream like crazy when you bite them."

"People say that about lobster, too," I tell him, edging my way into the attendant's closet. "Not about biting, but about dropping them in boiling water."

"Sorry?"

"Big saltwater crayfish."

"I know what a lobster is. Are you saying *you boil them alive*?"

"Personally, I don't eat any conscious creature. Many people love to eat them, though. They're a delicacy."

"Unfathomable depravity," says the turtle.

I turn on the light and find myself in a hallway that stretches in the opposite direction from the aquarium. "Climb out and bang on the glass," I say. "I need noise to find that access panel."

"If I could climb out, don't you think I would have by now? There's no land in here. Not a minute goes by that I do not long for the feeling of soft soil between my claws, for the taste of flowers and stick bugs and fish, for the sound of a thunderstorm and the taste of a fresh rain."

"There must be *some* way to make noise."

"You're saying I'm stupid?"

"I'm saying I can't find you without some help."

"Maybe you should judge us differently. Maybe someone's ability to figure out noisemaking isn't as important as his capacity to suffer. And believe me, I suffer, staring at that crapper all day."

I can't help but laugh.

"Something is funny?"

"That you know that word."

"Coined in memory of Thomas Crapper, like the word for shit."

"Actually, there was a Crapper Plumbing Company," I say mildly, "but the word crap is from the Latin for excess or chaff."

"That's fine, Mr. Monk-tionary. Argue and one-up me all you like. It's human nature to be unkind."

"Hey," I protest. "That's not fair."

"I'm talking about humanity as a whole."

I can't dispute the assertion when he puts it that way, so I return to the men's room and regard him once again through the glass. He's upright now, his feet in the gravel at the bottom of his tank, his plastron pressed against the glass so tightly that I can see the lines running in starburst patterns from his midline.

"Your shell is very beautiful," I say.

"Oh, stop. You make me feel so cheap."

"And the margins of your shell are so dramatic."

"Shut up and find me."

I try the closet again. I follow the hallway farther this time and find a spiral staircase I could swear wasn't there before. It's metal, like something one might find on a ship, with drainage holes and corrugations on the steps. I descend it slowly, step after step, my hand on the hard black metal railing. The lower I go, the more miasmal the air becomes. It gets harder and harder to breathe. Holding my breath, I continue my descent until I'm in a veritable fog, but one pleasantly redolent of flowers and comfortingly warm. My lungs aching and my heart pounding, I wonder if I've found the Garden of Eden and if the immortal has me, as I did at Olduvai Gorge, once again exploring the history of my own species.

I'm gasping, so air-hungry I'm about pass out and die, but when I reach the last step and see water before me, I can suddenly breathe again. I sit down, inhaling in greedy gulps, the stars behind my eyes receding, the world stabilizing, my heart slowing. The water glows in an inviting way that makes me eager to enter its primordial depths and discover its most intimate secrets. I remove my black-and-white kung fu slippers and slide in. The water is warm and welcoming. I submerge. My robes spread about me like a monochrome orchid, keeping me afloat, preceding me, following me, escorting me. The tips of my toes touch the bottom.

"Recognize where you are?" asks the pond turtle immortal, surfacing beside me.

Suddenly, I do. Impossibly, I'm inside his tank.

"You tricked me," I say. "Distorted my senses. Messed with my mind. Took advantage of my desire to save turtles."

"And you wished to be tricked," he says, his voice no longer plaintive and thin but now resonant and strong. "Some minds are like that. They can only access their true nature if they are beguiled, manipulated, and goaded into acting."

"So your pitiable complaints, arguments, and accusations were all contrived to help me see how far I would go to help?"

The immortal does a pretty good job of bringing his forelimbs together as if he is clapping his hands, something I've seen before. "No, just to remind you of what a human being can do if he really wants to."

"Well that was cruel. I almost died from lack of oxygen."

A female voice laughs. It's a beautiful sound, like a bell. I turn and see a radiant woman rising from the surface like the Lady of the Lake in the Arthurian legend. She graces me with a beatific smile. She is the bathroom attendant, but minus her rags and mop and bucket.

"I am Lady Yang Xi," she says.

"The mystic who brought the revelations of my *Shangqing* sect of Daoism to the Xu family back in the fourth century," I cry. "I'm honored! But how can you be here?"

"There's no rule that says we can't have company," the pond turtle puts in.

"Then who are you?" I ask, suddenly suspicious.

"It is as if I put you down a well and had you look up at the night sky to realize what you did not know," the turtle answers. "It is as if I made you a fish and then asked you to understand water."

"Zhuangzi," I exclaim, my chest bursting with joy. "The beloved Daoist writer!"

" 'Tis I," he laughs. "That inveterate spinner of tales, that stubborn, shining bubble of consciousness in the boiling water of your soul. In your eyes, I am a turtle. In truth, I am only that which comes from your own heart, your own longing, your own imagination."

"I didn't expect this," I say, feeling tears hot on my cheeks and reaching out so that Zhuangzi, strong and solid as he is, will steady me. "I didn't see where all this was going. I should have figured it out."

"Don't be so hard on yourself," he says. "You are, after all, the source of all you are experiencing here, the writer of your own play."

"The urinal, the aquarium, the pond. All of that was your imagination at work."

"Mine and yours together. Remember that nothing is as it seems, that your meditations are as real as the waking world in which you stand. After all, do you not find yourself as caring and moved in this world as you do by the people and creatures around you in that one?"

"I do," I say, suppressing a sob.

"And do you not have just as keen and clear a sense of yourself in this world as you do in that one?"

"Perhaps keener," I say. "Perhaps clearer."

"Does not your compassion have control of the tiller of your ship in this life at least as surely as it does in the world in which your hands are now folded over your navel, your eyes closed, your breathing imperceptible, your bowels, if I may say, roiling?"

"It does," I say.

"I ask you, then, will you help the rest of your kind see that it is the ability to suffer, and the desire to escape suffering, that is the thing that binds us far more surely than whether we have shells or scales or soft skin or hard, whether we have limbs or wings or feathers, cilia, antenna, compound eyes, or no eyes at all?"

"I will," I say.

And with that I am under the water and looking at the real and three-dimensional paradise I could only imagine when I was a boy paddling a canoe on a river in Connecticut and saw my first turtle. The plants undulate in the gentle current and the fish flash past. Crustaceans wave from below, and above me, on dropped and floating leaves, ants take voyages that feel a million leagues to them.

And, everywhere, absolutely everywhere, there are turtles.

THEY SO WANT US GONE

There are days I just can't stand still. It's a dirty little monk secret, but it's true. During first twenty years or so of meditation, I occasionally warred with the immobility required for so-called "Standing Pole" exercise, the art of making like a tree without leaving. At those times, I would meditate whilst walking. These days, even though I can reliably stay still, often for hours at a time, I still enjoy a good walking meditation session.

One reason it's so much fun is the variety of walking options available. Among these, one can move forward or backward at the same time as executing classical tai chi arm circles, one can "offer fruit" like an ape while stepping forward, walk backward flapping one's arms like a crane, or even slide one's hands up and down the front and rear centerlines of the body in imitation of a lumbering bear.

Today, because I've had too much tea and am not quite ready to settle down, I take a walk around the park. Even though I am in motion, my eyes are narrowed to slits, my breathing is slow and regular, and my attention is totally fixed upon my internal environment. I'm not likely to step on any glass or stroll into a fire hydrant, but otherwise, I am out of tune with the random, excessive, and often irrational vicissitudes of other people's behavior. The motorized skateboarder who nearly mows me down goes by without response from me, ditto the bicyclist up on one wheel listening to headphones, eyes closed. The water snake that slithers over my shoe is not booted up and away, and the naked lunatic twirling a black plastic contractor cleanup bag like a limp and flapping baton does not my equilibrium disturb.

It's business as usual here in the park, and I take advantage of trees for shade just as I always do, though there are moments where gaps in the canopy allow the sun a brief shot at my bald pate. As my mind begins to settle down, I find a straight tree surrounded by flat ground—no protruding roots to stumble over, thank you—and constrain my walking to a circle with the tree as the center around which I turn. This kind of walking, from a relatively recent style of kung fu known as Eight Trigram Palms (*Baguazhang*), emphasizes a shuffling, so-called "mud-slinging" step that disallows a raised heel or toe. As I circle the tree, I follow a sequence of eight different hands positions until, slowly, I stop moving completely, comfortable in my stillness, in tender agreement with the tree.

It is then that I notice I'm aloft again, with islands below, widely spread and covered in mist. I fly first above snow-capped mountains and then over ever-greener landscapes. Despite my high velocity, I see temples, factories, seaports, and cities, all dotted with the rising sun flag that tells me I'm looking at Japan. Then I'm over open water again and then over white sandy beaches and the lush landscape of Okinawa. I descend to an intimate distance. Avoiding the entangling underbrush, I follow the rise and fall of the terrain. Snakes slither beneath me, and the wingbeats of passing tropical birds are as loud in my ear as my own heartbeat.

I reach Naha, the prefectural capital. I see traffic and tall buildings and swarms of Japanese tourists. The natural environment is more developed than is good for any turtle, and I wonder what shelled reptile could survive it. I touch down on the side of a rural road that winds through a narrow pass, with steep hillsides on either side. It is raining heavily. Drops hit and cool me, drip off ferns and tropical plants, tug at my shoes in torrents, and cascade onto the road. Green frogs and tree frogs seek suitors in the storm, their calls piercingly loud.

A red SUV approaches. A young couple emerges—American teens wearing raingear and waterproof backpacks. They walk away from the car holding hands, pressed together as one, a bright yellow camera swinging from the girl's wrist.

"*Japonica* or bust!" the young man shouts.

His use of Latin tells me the pair is in search of the Okinawan leaf turtle, one of the world's rarest reptiles. Diminutive, terrestrial, and prone to hiding between rocks, it subsists on worms and other invertebrates and jungle fruits and mushrooms. It is highly illegal to catch this species, even more so to smuggle it out of Japan. The teens close in on my position. The rules of the game thus far tell me they can't see me, but, just to be safe, I duck down into a dark culvert that runs underneath the road. Its concrete ceiling is moldy and low, forcing me to squat. Algae grows on the stones scattered across the bottom. The collectors walk past, still excitedly talking about how much money they will make selling leaf turtles. I shake my head at their short-sightedness.

"Are you in a funk, Monk?" asks a tiny voice.

I squint to see a ridged brown shell, a beautifully red-striped head, and bold white eyes of a male leaf turtle. "You!" I say, astonished. "What are you doing in a culvert?"

"I live here. Though not for much longer as I'm starving to death."

I look at him more closely, and indeed I see the markers of chelonian ill health. His shell looks a bit depressed, his forelimbs are skinny, and the skin on his neck hangs too loosely.

"How did you end up in here?"

"A sad and ironic bit of murder that tale is. The Japanese government protects my kind strenuously. It's illegal to touch us at all, even if only to move us out of the roadway. At the same time, they build these spillways to minimize landslides and to protect the road in the event of an earthquake. Rain washes us in. We're good on rocks but we can't climb smooth concrete walls. Every time I try, I fall back down, often on my back. It's exhausting, especially as I grow weaker from hunger."

"Unintended slaughter," I say.

"Unintended torture. Starving to death is a terrible way to die. If it weren't for the *tengu*, we'd have other fates to worry about. Dying slowly alone in a culvert is a leaf turtle's worst nightmare."

"The *tengu*?" I repeat.

"Our guardian. A *kami*. A god of the forest. He looks after us."

"Good to know."

"Those kids will meet him soon."

"If this *tengu* takes such good care of you, why hasn't he pulled you out of this culvert?"

"He has done so twice. It's a big forest and he has many of us to watch over."

"What's he going to do to the kids?"

"Sooner or later, someone will report them missing, or maybe stumble across their bones while hiking. Should be a while, though. If there's one thing the *tengu* knows, it's where to hide bodies. He's been taking revenge for this forest for a thousand years. There are gods and demons, cryptids, and spirits, all through the natural world. Ghosts, too. They do the work that needs doing and people put them in poems and stories, songs, myths, and legends."

"I thought "cryptid" meant they were imaginary."

"That's what you think."

"May I have an example?"

"Sasquatch," the leaf turtle immortal declares without hesitation. "Also, the New Jersey Devil, the Everglades Swamp Ape, Big Foot, Yeti, and others."

"You're saying Bigfoot is a god?"

"Certainly. And most turtles wish to see more creatures like him and less like you."

"I grant you there are too many of us and that we do bad things, but we have our good sides, too," I say, feeling more than a little defensive.

The immortal blinks, wipes an eye with a forelimb, and shakes his head.

"Oh really? Like what?" he asks. "That you make music and art? That you send rockets to the moon? Believe me when I tell you that every thinking creature on this planet wishes to wake up one morning and be blessedly rid of your kind. Can you imagine that we turtles wouldn't be so much happier with our *tengu* able to tend the forest the way he used to before you took to burning and chopping and poisoning it, before you changed the rain to burn our skin and kill our frog brothers, before you soiled the waters to kill our friends, the fish? Don't you know we are tired of walking over the torn carcasses of birds chewed up by your jet engines? Oh, how we yearn to see the twinkle of countless stars in the night sky and the moon in its circular brightness rather than merely be teased by the nightly glow behind the clouds of pollution you create. How we dream of the sweet fragrance of wildflowers now masked by the stench of diesel exhaust. Do you really believe we want to hear your children scream and shout on ziplines above us, pouring foamy sugar drinks on our shells? Honestly, Monk, how could we prefer the damaged and chaotic world you have left us to the peaceful, tranquil harmony of the world before your kind? Of course turtles want you gone."

I collapse against the walls of the culvert like a crushed toadstool. "This lesson feels so bleak," I say at last and in a quiet voice. "Is there no hope for us?"

"Recognize you are nature's most malignant mistake and try to repair what you've done. That is your only path to redemption."

"At least I can save you from this jail," I say, proffering my hand so he can climb onto it, which he does.

"Thank you. I have a family in the forest. Is there a chance you can reunite me with them?"

"You guide, I follow," I say.

I am about to leave the spillway when he bites the palm of my hand.

"Ow!"

"Sorry, but I had to get your attention."

"You have my attention. You didn't have to bite. We've been talking right along. What kind of immortal is so violent?"

"You have to hide right now. The *tengu*'s coming! If he sees you holding me in your hand, he'll crush you. Believe me, he'll do it. He's fierce!"

Cautiously, I raise my head for a peek outside the culvert. The leaf turtle immortal bites me again.

"Hey. Stop doing that. Look, you made me bleed."

"I may be small, but I'm mighty. Now keep your head down. He's after those kids, not after you. Just be patient and keep still. I never saw such a hyperactive monk."

"All I did was crawl down here and talk with you. And fine, you're mighty. But really, don't bite me again. Those kids are just young and dumb. They may be poachers and deserve to be kicked out of the country, but I don't want to see anything terrible happen to them."

"So many excuses for the terrors human beings set loose upon this world. Anyway, the *tengu* doesn't care what you want. He's his own god."

An enormous creature, man-like, but of gargantuan proportions, flits into view. Despite being bright red, he melds with the dappled light of the forest as if he himself is made of shadow. This makes sense to me, as this meditative world I've conjured is itself some sort of shadow world and so a shadow of a shadow has great substance in the way that two minuses multiplied give a plus. His feet flash as he closes in on his oblivious, helpless quarry.

"I have to warn them," I hiss. "I'm a monk. I can't ignore murder."

"If he's on their trail, it is already too late."

"Hey!" I cry, tunneling my hands to throw my voice.

The *tengu* turns to show me a hideous red face and nose long enough for a dozen perching songbirds. His vine-like veins and grapefruit-size muscles ripple and bulge.

"Now he knows you're here," the leaf turtle declares, sounding vaguely satisfied. "He has preternatural sensitivity to all things in nature, all the things humans rarely take the time to notice, like the mosquito eggs in your drinking glass or the flies that watch you fornicate."

"I'm just trying to save the kids."

"Aha. Loyalty to your own wicked kind rather than compassion for their victims. What a flawed monk you are."

"Compassion is not species-specific," I say. "They deserve it, too."

Leaf turtle in hand, I vault up and out of the spillway and pursue the *tengu*, floating as much as running, my feet barely touching the ground. At close range, the giant god smells like a forest fire. The top of my head barely reaches his waist. The screaming teens dangle from his grip, swinging like clock pendulums, their eyes bulging, their fingers in claws.

"Be merciful, sir," I implore. My words come out of me in his ancient Japanese forest tongue, and somehow, I am not surprised. I'm also unsurprised he can see me. Rules of the game say humans can't see me, and the kids don't, but creatures of the mythic world know I'm here because I am one of them.

"Let one go, there will be a thousand more," the *tengu* roars in response. "They come only to plunder my forest."

"If you kill them, the forest will soon be overwhelmed by people looking for them."

"Not so overwhelmed. They are foreigners. They have no kin here."

Right then, he notices I'm still holding the turtle. He closes the distance between us in an instant. His pants are loose as a farmer's, but not so loose as to disguise his outsized manhood. His shirt is made of feathers. The breath of his hiss is so hot, it scorches the air between us. He's some kind of bipedal dragon, this long-nosed monster, this god.

"Put me down," the immortal whispers to me. "And do it gently."

I decide to take my chances with the red giant instead. "Wipe their memories then," I tell him. "A god like you must know how. If you didn't, you could never have survived killing people in this forest, in the era of machine guns and night-vision goggles and infrared heat detectors. I believe hikers see you all the time, but don't remember that they do. Tell me this isn't so."

"It's so," mutters the *tengu*.

"I rescued the turtle from the culvert, where he had fallen and was starving to death. I did a good deed. Please do one in return. Calm yourself and wipe their memories."

In a leap of faith, I lower the turtle to the ground. The *tengu* hesitates for a moment, but not more than a moment, then does the same with the kids. Instantly, they curl up and snore.

"Thank you," I say in that strange, long-dead tongue.

The turtle crawls away, the *tengu* follows him, and I fall into step behind them. We are, I am certain, one strange procession. The turtle looks back over his shell at me. He's getting harder to see as the whole jungle begins to fade. The last thing I glimpse I have before I am back by my tree in the park is a female leaf turtle, young in tow, scrambling toward her mate.

The *tengu* looks on.

Maybe he smiles.

THE MEDIUM IS NOT THE MESSAGE; THE MESSAGE IS THE MESSAGE

The park is closed today. I haven't encountered this before, perhaps because I am often traveling on holidays. Instead, I make my way to the nearby seashore. It's not so easy to find a quiet spot anywhere near where I live, as it is a densely populated urban environment. Still, there are some mangroves and palm trees between the parking lot and the ocean at my local beach and I discover a quiet, hidden spot that gives me an eye-level view of the ocean, puts my feet in the sand, and tucks my prying eyes so that I may pursue my practice undisturbed.

I rarely meditate barefoot. Perhaps that's because my meditation practice began, many years ago, as a facet of my tai chi practice and tai chi, having been conceived and developed in the cold, loess soil of northern China, is practiced shod. I enjoy this new sensation. My toes seem to have a mind of their own, not only enjoying the satisfying scratch of the sand, but wriggling down toward the liquid, bubbling molten center of the Earth as if in their own quest for energy.

While I'm employing the technique in quest of turtle trances, energy is what Daoist standing meditation is traditionally about. Specifically, it develops and strengthens the body's natural energy circulation—the Greater Heavenly Circle—which proceeds up the middle of the back along the Du channel, one of traditional Chinese medicine's so-called extraordinary vessels, passes over the top of the head and then down the middle of the front of the body along the Ren channel, another extraordinary vessel. This energy circuit takes place like the circulation of blood and lymph, like the circuit of the Earth about the sun, like the procession of days and nights. Absent training, most of us are not aware of it, but it happens anyway.

To stand and enhance the circulation of the Greater Heavenly Circle means to take one's proper place, suspended between heaven and Earth. It also means to connect to those energy sources, heaven through the acupoint at the skull's crown point, *Bai* Hui, also known as the Hundred Convergences point, or Du 20, and Earth through *Yong Quan*, also called the Bubbling Spring or Kidney 1, in the middle of the ball of each foot. These two points set, respectively, the upper and lower limits of the circle of energy. In another sense, achieving this proper place is to balance yin and yang in the body. It is also a description of humankind's spiritual place, lifted to heaven, but grounded to Earth. Melding all these concepts helps define what it is to be a person in what I feel is a singularly beautiful way.

I am soon in the world of immortal turtles, floating downward from on high. As I drop, it is my nose that tells me where I am. May I call it a fragrance tango, this thing that the Negev Desert and the Red Sea do to the air above the open water and the shoreline at Eilat, where in summer tourists cavort in the cool clear water under the watchful eyes of three nervous navies? Is it the ineffable qualities of this biblical wonderland that make me smile at the aroma, even though the last time I inhaled it I nearly drowned, swimming in cold water with a bum heart while children constructed sand castles, testosterone-rich boys warred on jet skis, and bikini-clad beauties paddled past on stand-up boards?

It is just before dawn, and the sun is gently kindling a fire on the horizon, setting a slight shimmer to the waves. Despite the tense, edgy, devil-may-care, we-might-be-bombed-at-any-moment-so-let's-live-life-to-the-fullest energy rising up from the landscape, I primarily feel excitement at the prospect of my next lesson. There are few species that live in Israel and even fewer that live any distance from fresh water, so I'm not surprised when I see a female Negev tortoise moving slowly across the northbound lane of Highway 90, the main highway serving Eilat. One of the smallest tortoises in the world, this one is barely the size of a camel chestnut, and, by the weathering of its shell, is also quite ancient. Just as I touch ground, she begins scratching at something at the edge of the tarmac.

"What are you doing?" I ask.

"Ah. It's you, Monk. I was told you have terrible timing."

"Nice to hear. What is that you're messing with?"

"The optical sensor for an Improvised Explosive Device," she says. "A few years back, they used pressure plates and trip wires, but those were sometimes revealed when the wind blew away the covering sand. From what they said as they installed it, this thing here is the safety sensor for a garage, the kind that stops the door from squashing the family cat. Ironic, don't you think? Meant to save lives, but here, used to take them?"

"You said they. Who do you mean?"

"You call them terrorists. They rigged it just before dawn. It's pointing in at the road, so when it goes off, shrapnel will penetrate any vehicle coming by."

"So it's a bomb?"

"Yes, Monk. It's a bomb."

We don't talk for a few minutes, because she's very focused on her task, and I am obviously inside the blast radius. I comment that

153

THE MEDIUM IS NOT THE MESSAGE; THE MESSAGE IS THE MESSAGE

messing around with a bomb might not be the best idea, but the immortal assures me this is the job she is here to do. She continues her work, pushing up the beam generator.

"If I move it enough, I'll break the circuit," she says.

"It sounds as if you're trying to set it off."

"That's precisely right."

"Can't we talk about this? There has to be a better option than blowing yourself up."

"There isn't. Sometimes self-sacrifice is the best way. Long Hair did it, after all."

"Long Hair? You mean Jesus Christ?"

"He didn't walk on water, by the way. Walked just like you do. The thing was the level of the sea was really low that day...."

"How could you have known Jesus? That's so long ago."

"What am I, Monk?"

"An immortal," I say slowly.

"So the answer's in the question, right?" the tortoise replies, circling the device, inspecting it more and more carefully.

"Of course. So, back to the question of blowing yourself up, what I'm trying to say is that we so often see things in binary fashion—by we, I don't just mean Daoists, I mean everybody—but actually, I find there is more nuance to the world than that."

"That's nice. But I don't recall asking your opinion."

"All right," I sigh. "I understand you're the one who is supposed to be giving the lesson here. I'm just saying that there's usually some third alternative to either doing or not doing something. In my own writings and teachings, I frame it by using doors as a metaphor. So door number one is using force against force. Door number two is yielding. Door number three is finding a creative solution unique to the situation. Like figuring out a way to neutralize this thing without blowing yourself up and me with you."

The tortoise immortal studies me for a moment. "Well, there weren't many doors in the land back then," he says, "except maybe in the Roman camps and the temples, and I frequented neither. Mind you, Long Hair had his opinions about the way things were and the way he wanted them to be, but whatever his delusions, impressions, or beliefs, he had only the good of all of us in mind. Treated us particularly kindly. We turtles, I mean, though he wasn't as keen on us as he was on doves. He never knowingly harmed a fly, though I did see his demeanor crack one time when a viper crossed his path. He stepped on it, put his sandal on its head, and was all set to crush it, but got hold of himself and backed off so the poor dear could slither away. This attitude he had about him, utterly crazy but so magnetic, it really pulled people in. One time, I saw a whole line of mesmerized creatures following him across the sand south of Galilee, though the viper had spread the word, so the snakes boycotted the parade."

"Were you around when he was born?" I ask, both because I can't resist the question and because I feel an urgent need to keep the Negev tortoise talking until I myself can divine a third option in regard to defusing the bomb.

"There was plenty of talk about that: about his mother, did she have a boyfriend, what kind of tent she kept. And just because I was camping nearby doesn't mean I was watching all the goings on. After all, how was I to know all your kind would end up thinking she was so special?"

"If there were goings on, you've answered my question."

THE MEDIUM IS NOT THE MESSAGE; THE MESSAGE IS THE MESSAGE

"Don't jump to conclusions. Things were different back then. No one held judgments, not even the stars, and the inside of a tent could be stifling to the most modest of women in summer. A female of your kind couldn't be blamed for forsaking modesty to indulge her natural urges out in the fresh air."

"I get the picture," I say.

A breeze comes up as the sun remembers its duty to its third child. The little tortoise extends her limbs and raises her head to the warming rays as all her kind do after a long, cool night. The pulse in her throat quickens, and blood flows to her limbs, causing them to tremble gently as her blood flows, her sinews soften, and her grateful muscles respond. White goo gathers in the corners of her eyes and salt appears at her nostrils.

"You're so curious, Monk," she says. "I wasn't prepared for that side of you. Nobody warned me."

"So who was her partner? To the devout, that's an important question."

"Actually, that's only a question for nonbelievers; for the devout, it's not a question at all."

"Are you going to tell me?"

"Do you see how badly you want to know, you for whom it's a purely academic exercise, you for whom the emphasis is on this life and not the next, you who prizes harmony and balance above all else, you who believes that you are a fractal little pearl made in the image of the cosmos?"

"You have me at a disadvantage," I say. "Seeing as how you know more about me, and what I hold dear, than I know of you."

"You know that immortals don't die. You know that we've seen everything and everyone. You know that we deliver messages across time and across cultures. You know that we take the form most appropriate for those to whom we speak. You know that we tailor the message for the eventual audience, the readers of spirit-writings and those who attend sermons. You know that we're not only Daoists, although for you we are, and you know we've talked to others and been other things."

This revelation, if that's the word for it, drops me to the sand. I suppose I realized all this on some intellectual level, particularly given that the Red-ear in my first experience had clearly said that immortals appear as they must to those who need them, and that scholars maintain that spirit-writers create their own teachers and, from a scientific point of view, are likely only to be listening to one part of themselves speaking to another part. Even so, to have this phenomenon laid out this way for me takes my breath away.

"I see this gives you pause," the tortoise says. "I sense you doubt the veracity of what's happening here. Don't do that. Be glad we deem your worthy."

"You have no idea how grateful I am," I say. "You have no idea how humbling this whole thing is."

"Well, even if you haven't entirely grown up, you want to, Monk. And even if you still cling to your sense of your walled-off self, you want to stop doing that, to see what unites us all."

"I've seen my parents sicken and die," I say. "I've seen my son gradually supplant me. I know what it is to feel alone sometimes and to come face to face with nothingness, and to realize that even after I'm gone, the larger world won't know or care. All those experiences are strong motivators in the pursuit of deeper understanding."

"Indeed, they are."

An Israeli troop carrier passes by at a distance, stops, disgorges some soldiers, and then continues on. I assume the soldiers are out on maneuvers as they are in uniforms of such effective desert camouflage that they disappear within moments, rifles and grenade launders and all. They are too far away to hail them, but I stand up anyway, figuring it's probably a good idea for them to know that I'm here.

"They can't see you," says the immortal.

I'd forgotten that. I take a deep breath. The desert heat has begun to gather, but the temperature is still comfortable and the fragrance of sand meeting sea still intoxicates me. I wonder whether it will persist into mid-day or whether all moisture, along with the molecules responsible for the smell, will be burned away by the sun. I wonder whether I myself will persist long enough to find out, given that I'm just a few meters from a bomb.

"I don't need to believe Long Hair's mother was visited by a supernatural being who fathered him," I say, realizing as I speak the words that they are a big *non sequitur.*

"And yet, I'm a supernatural being and you want to believe in me," the turtle replies.

"What you say is more important to me than who you are," I counter. "So as long as you help me stay innocent without denying what I know, find joy in my place in nature despite the way my kind is destroying it, and abandon guilt, guile, dissembling, and lies in favor of compassion for all that is, I'm content."

"Long Hair had many of the same ambitions and the same spiritual hunger, too. His language was more flowery, and he used more allegories, but the message isn't so very different. You know there are those who said he went east after he was brought down from where he was tied up. There are even those who say he is the source of all the Asian ideas to which you cleave."

"Timing's wrong for that," I answer. "He was five-hundred years too late to have added his two cents to the old texts I read."

"As if you can be sure about who wrote those and when."

"Are you telling me Jesus put you in his back pocket, took you to China, and brought you back?"

"He didn't come back. I had to find my own way on horses and camels following those winding silk roads, by proxy, of course, hiding in saddlebags and such, most of the time between hard tea cakes."

I cover my shock by suggesting that we interrupt the bomb trigger beam with a mirror.

"Tried it," says the tortoise, jutting her chin toward a piece of broken glass a few paces away. "I pushed that over here a few minutes before you arrived."

"I could try," I say.

"You think you're a better glass positioner than I am?"

"I don't know why I would be," I confess. "Although I do have fingers."

"If you make a mistake, you die," says the turtle. "I'm immortal, so that can't happen to me."

I smack myself on the forehead. "Duh. It doesn't matter if you blow yourself up."

The tortoise opens her mouth wide, showing her pink gums, then closes it again with a little click. "I wouldn't say that," she says. "Being physically destroyed always has consequences, even if we immortals do get to start all over again. I still feel pain, although exploding happens too fast to hurt much, and I still leave this realm and end up somewhere else."

"My meditation teacher taught me that to meditate is to link to heaven and Earth and thereby deconstruct myself," I say. "He told me that the descent into chaos we see these days is merely part of the universe expanding before it once again contracts, and that if we lose everything, that's all right, because those who come after us will likely have just what they need when they need it."

"A very Daoist description."

"But not something Long Hair would say, right?"

"Not exactly. I'd ready yourself, Monk. The school bus will be here soon."

"What school bus?"

"Don't you know? This is a historically important spot we're occupying. It's north of Eilat and south of Be'er Ora. In 1968, Palestinian terrorists bombed a bus full of high school students there. That kicked off a retaliatory battle. History has a way of repeating itself. Children still have to get to school in town even though the new airport stands where the old village once was. I've got to set off this bomb before they get here."

"Why don't you let me wave down the bus?" I say, "or call in the experts to defuse the bomb?"

She turns to look at me, a gentle and patient look in her eyes. ""We've been over this. You're not here for anyone but me, remember?"

The light is growing, and I get a better view of her subtly beautiful two-toned carapace, the dark starbursts in her vertebral scutes, the camouflaging shade of her background coloration, her tiny dimensions, the light patch on the hard plate atop her head between her eyes. I notice that one of those eyes is rheumy with a cataract and wonder how good her depth perception can be as she works with wires and clips.

"But this is terrible."

"Letting the children die would be terrible. All they have is the days of their lives. We are not, in fact, creatures of space and place, but of hours. So many human beings forget that and trade their most valuable and nonrenewable resource, their time, for renewable resources like money."

People often think the desert is flat, and, compared to mountains, it may well be, but there is topography to all of Planet Earth's bone-dry climes, from the famous dunes of the Sahara to the craggy ravines of Sonora. This road through the Negev follows the rises and dips of the land atop which it floats, and there is thus a moment when the bus is visible in the distance and another moment when it is lost to sight. Even so, I know it is coming and that the tortoise cannot be dissuaded. She will not wait to set off the bomb until the bus is close but will instead detonate it when they are still at a safe distance. There isn't much time left.

"In all this, you haven't once mentioned good and evil," I say suddenly. "You haven't once mentioned the terrible people doing this terrible thing."

"Long Hair would have harped on that," the immortal answers. "I'm not inclined to do so. So much of what we deem good and evil is a matter of perspective, of our beliefs, of how we see what is happening, how we frame our actions. If you believe that sending these children to heaven because they are heathens is a blessing, and if you believe you will be rewarded for accomplishing that blessed act, then you are not committing evil, are you?"

I worry about getting drawn into such a debate, particularly with the bus fast approaching. "I see it biologically," I say. "Good and evil are about power. When the powerful do something against the will of the powerless, that's evil."

"You say," says the immortal.

"Tell me you don't agree! You must, or you wouldn't be sacrificing your life here, even if you do get another one right away!"

"I don't agree because I don't bother with such thinking. The bombers do what they do; I do what I do."

Exasperated, and remembering that I came here by floating downward, I float up, rising into the air. I fly like an eagle after a mouse, low over the landscape, hands extended like talons. I see something I think might serve to reflect the beam back to its source and thus deactivate the switch, so I dive for it. It is a knife, not exactly rusted—this is the desert, after all—but dulled by the scouring winds of Israel. It is curved like a scimitar but too small for a sword, and it is sharpened on both sides like a dagger. Perhaps it is of Arab design, perhaps it does not belong here at all, but was dropped and lost by a conqueror passing through, an assassin, mercenary, crusader, or thug. I grab it and fly back.

"Here," I say breathlessly. "What about this?"

"Too late," she says.

The bus is close enough now that I can see it is full of little heads bobbing to music I can barely hear, the choice of the driver streaming backward for the kids. It sounds Moroccan, but maybe it is Malian or even the driver's cousin's bar band from Tel Aviv. A couple of small brown arms dangle from the windows. The bus is made by Blue Bird, which I happen to know is a Fort Valley, Georgia, company. I wonder how it got here.

The tortoise takes one final bite and makes it through the wire. Her body is the conductor now, all those salts and cellular juices, all that interstitial fluid, lymph, and blood maintaining the circuit. I go down to ground level so that she can take me in with her one good eye.

"You're sure?" I ask.

"If I wait another minute, I and the children all go," she says in a wire-garbled voice. "If I do this now, they live."

And with that she releases the wire from her mouth, interrupting the circuit, and sending that one final rush of electrons to the detonator.

I am suddenly in the midst of a new, small sun, one whose life is measured in milliseconds rather than eons. It arises, it makes heat, and it fades. All around me there is light and there are a thousand flashes that birth a thousand memories. I wonder whether this is what death will look like when it finally comes, not a tunnel with a glow at the end, and not a film reel of experiences, but rather a firework display, each pop a snapshot of a landscape, a lover, a situation, a thrill. When the light subsides, I am already floating upward. Below, the cracked road shows no sign of a tiny turtle, not even a hint of shell or bone, for all those have been vaporized back to Dao.

It does, however, show a school bus skewed slightly sideways across the road, trailing tire marks but safe, upright, intact.

Back at my beachfront park, I see children squealing with delight as they run in and out of the surf.

It's such a sunny, beautiful day.

BALANCE COWARDLY ACTS OF EVIL WITH FEARLESS ACTS OF BEAUTY

There is no way to be outside today. The wind is up, the raindrops are the size of a baby's fist and coming fast and hard, and the grounds of the park are flooded. Instead, I stand in my meditation room at home, in classic Daoist posture, my arms outstretched as if hugging a tree. My robes rustle in a slight breeze coming through the window. Pinon incense burning by the door helps me to get in the mood for a turtle journey. Chinese flute music presents an eerie curtain to the sounds of neighboring leaf blowers and lawn mowers, the snarl of passing motorbikes, and the incessant barking of my neighbor's penned dogs.

I close my eyes. I inhale and exhale rhythmically, lightly, gently, moving the gas exchange from the surface of my lungs to my skin. My respiration becomes so light and slow, it is barely perceptible. I manipulate my energy in the very specific way these trances require. I feel the transition to turtle time approaching, a wispy, weightless, delicious sensation. I smile in anticipation of the light, zooming feeling I associate with the impending wisdom of another immortal mentor.

What I get instead is the slap of jungle leaves in my face and the scratch of branches against my arms. I feel a pressure in my chest, try to relieve it with deep, gasping breaths, and find the air so heavy I might as well be underwater. Light filters in. I feel something on my leg and look down to see a giant blue centipede investigating the spot where my white monk leggings meet my sock. I brush it off before it bites me, see a large scorpion on my foot, and shake that off too. I press on. It gets drier. I feel a change in temperature and humidity out of synch with the progress I'm making walking. I accept it, for this is a place where only internal logic obtains.

The jungle eases, thins out, releasing its hold on me. Branches no longer snag my legs, my robes flow without restriction through the landscape, and a dry breeze comforts my bald head. I glimpse pieces of horizon through the foliage and press on toward it until suddenly, and without warning, I break out onto a vast lawn not recently tended or mowed. Before me stands a castle. Truly, there is no other word for it. Candy-colored spires reach for the sky like pink, tantalizing treats for the gods. Perfectly blue swimming pools dot the grounds betwixt its pavilions, some long and narrow for lap swimming, some broad and shallow and safe for children. Between them, arched breezeways bursting with bananas, papayas, and passionflower vines create green cloisters lush enough to please the most ardent of hedonists. A chapel with a cross atop its peaked roof lies at the heart of the quadrangle formed by the walkways.

This is clearly a luxury hotel, and confident expectation of business has been injected into every brick, window, roof tile, and door. All the same, rather than sunbathers, golfers, or staff, I see pervasive signs of unfinished construction: patches of incomplete tilework, barricades, ladders, paint buckets, patching pans, the smell of glue and turpentine and gasoline, too. I conclude it must be Sunday or a holiday, since even security guards seem absent. I wonder if there is a worker's strike. Drawn to the chapel, I find a pond in front of it and beside the pond, a bench. I sit and gaze at water choked with floating plants. The air is still and silent, smelling softly of salt. It is a peaceful and beautiful moment. I brush the jungle from my robes and admire the light reflecting off the pink towers.

BALANCE COWARDLY ACTS OF EVIL WITH FEARLESS ACTS OF BEAUTY

Movement in the pond catches my eye. A small female turtle emerges from the weeds, puts her forelimbs on the concrete edging, stretches her neck the way turtles do to facilitate a climb, and hauls herself out of the water. She is about five inches long with a low yellow-green shell, a robust beak, and all but absent plastron. Her very distinct appearance tells me she is a narrow-bridged musk turtle, and if that is the case, this must be Mexico.

"Greetings, Monk," she says.

"And to you, Immortal. Tell me, where exactly are we?"

"The Gulf side of the Yucatan Peninsula," she answers in a husky voice, startling in its overt seductiveness and a jarring contrast to her decidedly unsexy shape. "State of Campeche, near Ciudad del Carmen."

"And this..." I gesture at the buildings around us.

"A development project gone wrong. Financed by a *narcotraficante*."

"A drug lord?"

"He got shot last month. Didn't pay the police enough, apparently."

"And the building just stopped?"

"Oh yes," she says, crawling to her left a little bit so she can enter a patch of sunlight. "Shame, too. They were so close to opening. Of course, there are a lot of over-the-top resorts on the peninsula, so nobody will miss this one. No legitimate investor will touch it for fear of being associated with the cartels, and no cartel member will wear another's soiled underwear. Anyway, we can't mourn the failure of a project conceived and funded to benefit a murdering monster."

"It's hardly soiled underwear," I say. "This place is magnificent."

"The jungle will reclaim it soon. Look at this little pool and you can see that process already underway. Soon enough, it will be like a lost temple, all overgrown, peaceful, natural, quiet."

"Sounds like that's what you're hoping for."

"Cycles," the musk turtle says. "Jungles come and jungles go. Of course, before humans showed up, the cycles were much longer. There's a hyper, frenetic quality to everything you do, and it spills over onto the world around you. I've been squeezed out of my natural habitat by all the buildings, the roads, the airstrips, the dumps and landfills and poisoned ponds and streams. This is the perfect place for a turtle right now, as the pendulum swings back the other way."

"The crazy pace you mention? I think it's because we're the only creatures that understand our own mortality. Since we know we are born only to get old, get sick, and die, we try desperately to fit everything in before we go."

"A vain and fruitless strategy that only leaves you all feeling hollow, unfulfilled, and frustrated."

"Perhaps so," I say.

A butterfly lands on the musk turtle's head. The turtle makes no attempt to snap at it, despite the fearsome reputation of its particular kind, but simply waits patiently for it to depart.

"Do butterflies taste bad?" I ask.

"On the contrary, other turtles tell me they are delicious."

"So you're not hungry?"

"I'm starving. The groundskeeper stopped cleaning the pond a month ago, but even so no frogs have come, no crustaceans have appeared, not even any crickets. There's a whole wall of killing chemicals around this place, the result of constant spraying. It's deep in the soil and would be in the water, too, if there wasn't a filter still running."

"So why not eat the butterfly? Is it tainted, too?"

The butterfly flaps his wings angrily as if I've offended it.

"Butterflies have souls," clucks the immortal. "Frogs, newts, fish, crawling bugs, crabs, worms, snails—I'll go for those. When they die, they die. Butterflies are different."

"I have a feeling this lesson is about cycles," I say. "Are you worried that if you eat the butterfly it will come back and haunt you, life after life?"

"You're missing the point," the turtle says, still motionless for fear of disturbing the butterfly.

"Before you *tell* me the point, would you let me know why you think the butterfly chose your nose as a place to rest? Do you think it somehow knows your beliefs and therefore feels safe with you?"

"You're so silly. He doesn't know anything. He landed on me because the ground is too low and those trees are too high."

There's the chelonian practicality I've come to expect, the cut-through-the-BS straight talk so characteristic of my shelled friends. "So then, what's the big point you mentioned?"

"The vast majority of what you naked apes know has been told to you by someone else. That means you really don't know what you think you know. You believe credulously and without much circumspection. As a species, you resemble sheep far more than you resemble most monkeys I've met, though I can't say I've met other so-called great apes. Most of you are completely divorced from direct experience, swallowing world views created by others to enslave and pacify you."

I think about the history of Mexico, the human sacrifices, the vast empires built upon highly stratified societies, the gods with the beaks and wings of birds, the rainbow of infatuations and delusions lain upon the Aztec and Mayan peoples in their desire to conquer, suppress, and find meaning in life. I wonder how much and how many of those delusions have crept into the regular turtles of Mexico, not this oh-so-special immortal. I wonder how affected they have been by the food they eat and the water they drink. I wonder how much has entered and possessed them through the air they breathe. I wonder how much of their shells and bones and skin and brains are comprised of the same molecules that fell into the dirt when human babies were thrown into sacrificial pits. I wonder if Mexico's turtles are now, or can ever be, free of the energetic forces of the land that compel anything and everything that arises from it. I wonder the same thing about human beings.

"What can any of us do about that?" I ask, taking a seat on the concrete edge of the pond. "We're all products of our environment."

"You can wake up. You tear down all that you have built that imprisons you, all the ideas that limit you, you choose to see the choices that you have made that lead to your unhappiness, you let the scales fall from your eyes, as the Western Bible suggests in Acts 9:18."

"I'm not surprised you know the Bible, this being a Catholic country."

She retracts her head. The butterfly flutters away. An updraft catches it, and suddenly, it is shooting upward like an Apollo booster rocket. Toward the pink spires it goes, and I watch it, shading my eyes with my hands as it melds with the cotton of the clouds, a mere mote in my eye.

"Actually, it's a Mayan country," the immortal tells me. "And an Aztec country. A Toltec country. A land of shamans and healers, of seers and magicians and tyrants and thugs. It's the territory of the most exquisite females of your kind and the most passionate males. It is home to the wise and the stupid and the rich and the poor."

I rise and stretch. Looking around, I see a groundskeeper, the first sign of human activity since I arrived. He's in green overalls, uncomfortably warm in a climate like this, but perhaps right for someone in intimate contact with cacti and stinging tendrils under a scalding sun. He pushes a wheelbarrow with one hand and balances both a shovel and a rake on his shoulder with the other. He's got a wide-brimmed hat with a dark sweat line above the brow. He moves steadily in the direction of the pond.

"Is that groundskeeper the only person working here now?"

"He is. And he's coming to kill me," says the musk turtle. "I chose this place to meet you even though I knew that would happen and that I would have to suffer. It's been worth it, honestly. It's an exceptionally beautiful place."

"What makes you say he's after you?"

"He's got chlorine in his cart. He's going to dump gallons of it into the pond to kill all the plants and clear it out. I can't survive that."

I crane to see the contents of the wheelbarrow, but the groundskeeper is still too far away.

"Why don't you climb out of the pond?"

"I didn't know when he was coming. I was hungry and came to the pond looking for food."

"Go now, then."

"It's too late. He'll see me, and who knows what he'll do?"

"So you accept death by chemical burns? Why are so many of my lessons with you immortals happening when it's either late in the game or too late?"

"Don't you know?"

"Because that's the plight of turtles in the world? Of people, too?"

"Bingo," says the immortal.

The groundskeeper stops on the far side of the pond, perfectly aligned between the center of my bench, the midline of the water, and the door to the chapel. He sets his white plastic bottle of chlorine on the edge of the pond. I approach him. "Don't put that shit in the water," I say.

"You know very well he can't see you."

"Worth a try anyway," I answer.

"Acceptance is not your strong suit."

"I always try everything before accepting the unacceptable."

Looking right through me, the groundskeeper waves his hand in front of his face as if brushing off a mosquito. I step close enough to breathe on his face. He rubs his chin, looks around. I take hold of his hands like a preacher imploring a parishioner. I feel his flesh, but he doesn't feel mine.

"It's not only this pond," the immortal says conversationally. "A lot of the water near the big resorts is polluted with offal. This is actually better than where I came from. Are you not familiar with John 5:4?"

"The Bible again?"

"For an angel went down at a certain season into the pool and troubled the water: whosoever then first after the troubling of the water stepped in was made whole of whatsoever disease he had."

I throw up my hands and watch as the groundskeeper continues to rake, humming to himself, then takes a broom and scrubs the sides of the pond, turning the previously clear water turbid and green.

"Where exactly have you gone?" I ask the turtle, half hoping the answer is that she has been raked up and is hiding in the pile of weeds.

"I'm hiding on the bottom."

I look around in desperation, trying to figure out a way to stop the groundskeeper from using his bottles. I spy a pack of matches on the ground and an idea comes to me. I bend over and pick them up. The groundskeeper doesn't notice their apparent levitation.

"The hotel is totally empty?" I ask.

"Has been for a while."

"And you're sure it was built on drug money and that it will profit *narcotraficantes*?"

"Of course, I'm sure. Sooner or later another one will take it over. No legitimate buyer would dare come near it."

I walk over to the chapel, see the air-conditioning unit in the alley beside it. There is also a water heater presumably for the restroom. A pipe comes out of the ground and goes into the heater. I loosen the pipe until I smell natural gas, run away, and toss the match over my shoulder. Matches don't fly well, so it falls short. I throw another one. This one lands, and the explosion is far more spectacular than I expected. There is a whooshing sound, a rumble, and then a fireball that engulfs one side of the building. The groundskeeper rushes away from the pool, not toward the fire, but toward the main entrance of the hotel, presumably in search of aid. The fire spreads rapidly. Waves of heat and the acrid stench of burning wood, the pews, I imagine, pour out of the building. The gas continues to seep from the pipe, feeding the fire. I run back to the pond.

"Come out," I cry. "Your heaven is burning!"

Explosions echo behind us. A pink spire collapses with a crash. The fire roars. Approaching sirens wail. I grab the turtle and float up into the air. The immortal starts to laugh.

"Now I know why they chose you," she cackles. "You're outrageous. People will see me."

"Of course, they won't," I say as I head for the green mountains in the distance. "You're just a speck in the sky now."

We pass parking lots and factories. We pass countless honkytonk bars, and we pass tourists weaving in and out of traffic on little scooters. We pass pickup trucks so full of workers their rear bumpers create sparks on the tarmac. I glance over my shoulder and see a column of smoke rising from black-streaked towers. Speedboats zoom through the water offshore, leaving trails of gas and oil. Airliners descend and others climb. Trucks bring food and supplies to hotels. Vans deliver laundered sheets and towels. A Cessna pulls a banner advertising a domestic beer. We fly on. At long last, we arrive at a narrow river surrounded by deep jungle. There is no development anywhere in sight.

"You really did this," the immortal breathes as I descend.

"It's my job," I say.

"Good. Now go write about it," she tells me, and with that, I am back in my meditation room, breathing incense and listening to the sounds of my home.

PERSPECTIVE IS A TOOL FOR US TO USE AS WE CHOOSE

When I arrive at my meditation park today, it feels much smaller than usual. I can't really explain it, but it just feels as if someone has shot it with a shrinking ray, as if everything is closer together, as if even the sky is lower and the lake shallower and each and every blade of grass and tree is 20 percent less tall and wide than before. I ascribe this to some constraint in my own thinking. It's as if I'm not the monk I have become, but am instead burdened by something from the past, something near and dear and important to me, yet something keeping me back and down, something that won't let me grow. I shuffle around, confused by the feeling. As I settle down, my meditation feels different. Unusual. There is a presence around me, something insubstantial as filaments of spirit from a dream-spider's spinneret, yet somehow familiar.

"Dad?" I say.

"Dear Boy," he answers, as he often did when he was alive.

"How did you find this place?"

"I was thinking of you and your turtles. The intersection seems to have some special power."

I'm surprised. In all the years I kept turtles as a child, he never paid them a moment's mind. My second thought is to wonder where he is.

"Are you in heaven?"

"Don't ask," he says.

"Are you in hell?"

"How could you suggest such a thing?"

"Where, then?"

"It's not what you think. Everything's different."

He seems strongly disinclined to elaborate, so I move on. "I'm learning from the turtles," I say, sounding and feeling like a little boy again. "A particular one in Japan told me that the world is full of gods and demons, spirits and ghosts."

"Wise animal."

"Are you one? A ghost, I mean."

"What do you think?"

"You're dead. I guess that means you're something like that."

"Well then."

"You always told me to be and do the highest, best I could. For me, that means discovering all I can about what lies below and beyond and above the waking, solid world. I want to be wise, Dad. I don't want to waste my time on a lot of the nonsense I see others do."

"Perfect," my father says. "Keep doing that."

I cry as he disappears. A moment later, I land hard on the sandy ground of a tropical island paradise, a coral atoll of breathtaking beauty with mushroom-like structures rising from the sea and low foliage and perfectly clear water. I collect myself, bend to shed my

PERSPECTIVE IS A TOOL FOR US TO USE AS WE CHOOSE

shoes, and begin walking the beach. After a minute or two, I spy a male giant tortoise lumbering my way.

"Welcome to Aldabra, Monk," says the immortal tortoise.

"We're in the Seychelles, then."

"Yes, indeed."

"It's beautiful here."

"We do love our mangrove mudflats, our seagrass beds, our sunsets, our lack of predators, our solitude."

"Not as many tourists here as I've seen gawking at your giant cousins on the Galapagos."

"Galapagos are individualists," the giant tortoise sniffs. "As such, we disdain them. Couldn't decide on a shape. Got old Darwin in a tizzy. He should have come to the Indian Ocean instead if he was looking for real turtles."

"He did," I say. "Although, if I remember it right, he didn't go much farther north than the middle of Madagascar, Mauritius, and Rodrigues Islands. Not so far north as this, so he never saw your kind of giant tortoise."

"Had he reached Aldabra, he would have understood more about altruism and community, concepts he wrote about, but so far as We are concerned, did not deeply understand."

"He regarded altruism as a nearly fatal challenge to his theory of natural selection," I say, sitting down in the sand and letting the giant shell beside me offer some refuge from the blazing sun.

"Just so," the giant replies, his voice melodious, soothing, almost hypnotic. "And yet the irony there is that it is so adaptive, even ants practice it."

"Yet insect societies have castes—workers and warrior and royalty and such."

"And such indeed. Aldabras are a more egalitarian crew. Of course, it is easier to be so than it is on the aforementioned barren, volcanic wasteland."

"I've been to the Galapagos," I say. "I've seen scores of tortoises sharing mud baths there..."

"And have you seen the mating rituals of those cultureless beasts, monk?"

"I have," I allow. "They're a bit loud, I grant you, with all that bellowing and grunting, but I don't see the Galapagos as barren."

"They're trendy. He who falls prey to trends is a sucker."

"If I'm a sucker for anything, it's turtles."

"We are all plagued by squatters in our minds. It can be difficult not to trip over them."

An electric-blue coconut crab approaches my bare toes, which protrude from my black leggings. I know this is the largest land crab in the world, and I wonder what magic there is in these islands, something in the soil, perhaps, which gives birth to an abundance of size in many things. His claws would shame a pair of Vice Grips. I strategically withdraw my feet as he waves them, his stalked eyes drilling me for depriving him of a tasty meal.

"Give him a coconut," the giant suggests. "If you don't, he'll dog your steps until the end of your days."

I spy a coconut palm and shake it. Nothing falls. Shimmying up the tree in robes seems unwise.

"One of those will do," the tortoise says, pointing his nose.

PERSPECTIVE IS A TOOL FOR US TO USE AS WE CHOOSE

I see the coconut and collect it. The crab trundles toward me, blue claws snapping. A female Aldabra giant tortoise emerges from the bush and prostrates herself before the big male.

"My King," she murmurs.

"Wait," I say, picking up the fruit. "You're king of the Aldabra tortoises?"

"And of the green turtles and the hawksbills that nest here as well as the rest of the island. The rest of the archipelago, too, with all its rare species, including the Inaccessible Island rail, the smallest flightless bird left in a world decimated by your kind."

"I had no idea."

"That does not say much for your observational skills, Monk. Have all turtles you encountered referred to themselves in plural? Should We construe that my natural grandeur, speech, stature, and bearing do not impress you? Should We feel insulted?"

"I guess I should have noticed the difference," I say. "I mean, you're already immortal, so why would you care about being a king?"

"What a strange question. We step into the shoes of these lessons just the way you will step back into yours when you're finished walking on the beach."

"You mean you circle, immaterially, until the right moment to incorporate?"

"Something like that," he smiles.

I think of my father. I struggle to understand the nuances and complexities of the life and times of the spirit-writer, of the job I have to do and all the different ways in which my literary material comes to me. I conclude that it is my job to discern as much as I can so as to understand the lesson, but not to judge.

"Great," I say. "Got it."

"Now that you do, you may address us as Sire, or Your Majesty."

"Very good, Sire."

"Tell me, Monk—can you reconcile your Daoist beliefs with the fact that everything in our lives, whether we are entranced or assuming the extreme risk of eating stonefish meat at the sushi bar, is just a fancy videogame with no reality or substance?"

"I'm not sure I can believe that," I say weakly.

"The many worlds hypothesis demands this conclusion," His Majesty replies. "The odds are simply stacked against us living in the One True World when that world is in the company of an overwhelming number of simulacra. We wouldst like to imagine that so sublimely intricate a creation is admired and treasured by sentient beings across the galaxies. Often, at night in the cloudless skies above my islands, We observe the twinkling of stars and imagine they are the appreciative eyes of such gamers."

I toss the coconut at the crab. It rolls to a stop and he is upon it at once, embracing it like a lover, prying into the husk for the reward of the sweet white meat. Frigate birds, the gliding geniuses of the bird world, wheel overhead looking for other birds to harass out of a meal. A snow-white fairy tern, elegant and beautiful in the extreme, goose-steps past, rummaging through the coconut fronds. A bright-red Madagascar fody the size of a finch flits through the mangrove branches, lands briefly upon the shell of the Aldabra king, and promptly disappears. A lesser noddy stares at me sagaciously from a branch nearby, while a whimbrel plays a game of chicken with the surf.

PERSPECTIVE IS A TOOL FOR US TO USE AS WE CHOOSE

""A nice image," I admit, "but still difficult for me to believe."

"You feeling that way is a feature of the program. I'm here to help you see past it."

I know I collapse a little bit at that. It's such a bitter pill to swallow.

The king opens his maw, showing its pink interior, then closes it again with a snap. I notice the fine serrations on his beak, the better to sever *Opuntia* cactus, which I have read are the favored food of this species. I can tell he sees my discomfort and wants to change the subject.

"Once, in the Pleistocene, there was a giant crocodile here. He feasted upon my forebears," he tells me.

"You realize that when you talk about Aldabra Island through the span of geologic time, you are both crediting the designers of this game with a truly long-term vision and the players with incredible staying power."

His Immortal Majesty places one forelimb after the other upon a fallen coconut trunk, raising himself so that he towers over me, though, in truth, I am still seated in the sand. "You conflate our sense of time with those of the makers," he says. "Do you not see the failure in logic there, the assumption that our own perception of time is anything other than a feature of the game, a quality built into us in the same way as our very form? Who is to say that what passes here as ten millennia is, by the lights of the makers and in the framework of the game, anything more than the blink of an alien eye, and an aeon but an hour?"

"I suppose that's possible," I answer, slowly realizing why this great fellow has been crowned sovereign of the archipelago.

"The creators of the game could, of course, be long gone. They could be the ancestors of the current players. If so, we should admire their ancestral handiwork, bow to them as superior to us in the way that we are to ants, and be grateful we have a chance for this experience."

"In China, there is a bit more appreciation for ancestors than there is in the West," I say. "Even so, if what you say is true, I bow."

"Or it could be our own descendants who have conjured us—that *they* are the makers."

"I suppose it makes sense that some mixture of curiosity and respect about their own past might lead our descendants to bring us to life in this way."

"Even what we call the laws of physics, the rules of nature, might be no more than the sophisticated code used by our distant offspring in writing this program," His Immortal Majesty offers, spying a flat leaf of his favorite cactus and biting off a chunk the size of my head.

"All this is at least a reason to live an interesting life," I say. "I mean, if we're regular and boring, we're likely to end up as computer wallpaper or pawns who are sacrificed early to some higher strategy. If we keep doing weird stuff, writing boundary-pushing books, for example, or ruling over an archipelago, then we are less likely to be the character deleted by meeting some grisly end."

"Like the tortoises of the Galapagos," His Majesty says triumphantly enough; it is clear he is proud of his own logic. "Did not the grand majority of those boring, conforming, self-congratulatory beasts suffer ignominious ends, living food stashed into the hulls of the maritime vessels that bore whalers to their cruel sport? Did not the various races on the various islands of that inferior archipelago not largely die out while those under my purview have thrived?"

"The geography is different," I say. "I mean, more people visit those islands, not that I wish to take anything from the magnificent job you've done as sovereign ruler, keeping destructive humans at bay.

PERSPECTIVE IS A TOOL FOR US TO USE AS WE CHOOSE

The Galapagos do, after all, sit right in the middle of whale waters and not so very far off the coast of South America..."

"We are closer to Africa than those pretenders are to their continent."

"Yet those waters were closer for European whalers. There was no Suez Canal back then, so they would have had to sail clear down the west coast of Africa and around the horn and up..."

"This, too, has been programmed," says the king. "Though the Seychelles are becoming more and more crowded with tourists by the day."

"If I understand your lesson correctly, then to play by the rules is the best we can do, actually the *only* thing we can do. If that's the case, then why bring up the subject at all? Why even have the lesson?"

The king swallows, issues a belch, breaks wind, and passes enough chalky urine to fill a large salad bowl. "Perhaps the other immortals have judged you harshly. In truth, you are not the very stupidest monk. The answer, of course, is that knowing we're in a game gives you perspective. You can still enjoy it as the greatest possible game you could ever possibly play, but the downsides to it have less bite when you know it's not as real as it seems to be."

I think about this for a while. "It seems cynical to me," I say. "Like spinning a vinyl record in a vertical plane, like a bicycle wheel, when there is no needle there and no music to come out."

"Up to you," says the immortal king. "Perspective is always a tool for you to use as you see fit."

Disquieted by the whole conversation, I try one more argument.

"I've heard scientists say that the complexity of a simulation like the one you propose increases exponentially as we add in variables. The degree of complexity in our world would be impossible to manufacture artificially, even with an unimaginably powerful machine."

"What is impossible?" His Highness counters. "Perhaps we are designed not to be able to imagine such a machine. What's more, why should the laws of physics inside the game be the same as those outside it?"

I collapse on the sand, outplayed. "Your Majesty, I have to admit I have never met an immortal turtle like you."

"Such are the rules of the game, Monk," he says, doing that always-awkward turtle approximation of a smile. "In this simulation, there is only room for one king."

I stare out at the beautiful flat sea. I imagine Arab and Austronesian traders landing upon these delicate, graceful shores centuries ago after carefully negotiating the surrounding reefs. I imagine them first laying eyes on these beautiful, black, stalwart, and insightful giants. I wonder if any of them knew the king when they saw him, even though he might have been but a tyke at the time.

A tyke against whom all arguments are futile.

The island fades as my gaze goes glassy. I hope to sense my father again during my return trip and tell him what I have learned.

Sadly, he is nowhere to be found, and I am back at my park, where everything has now regained its normal size.

PERSPECTIVE IS A TOOL FOR US TO USE AS WE CHOOSE

SEIZE THE DAY IN YOUR OWN SPECIAL WAY

When the jet stream currents are just right, storms in the Sahara send dust as far as my home in the Southern USA. This makes for spectacular sunsets, as the tiny particles refract and reflect the declining rays of the sun like countless tiny mirrors, conjuring fresh colors, shades, and hues in true Impressionist fashion. One of those storms has reached my park, and I'm here at sunrise this morning because what is true of sunset is true of sunrise, as well. The dust tickles the back of my throat and makes my lungs ache, but the glorious sky makes that minor irritation worth it.

During meditation, I nearly always close my eyes all the way. This time, however, I am so reluctant to lose sight of the beautiful sky that I begin my session with my lids set to slits. This creates a warm feeling, like standing before a roaring fire, turning the inside of my head a comforting orange. Dust can cause coughing, which is not so good for meditating, so I breathe very carefully, taking shallow sips of air through my nose and relaxing all my muscles so that my oxygen demands are as low as can be. As before, once my breathing is shallow enough, it feels as if I'm breathing through my skin, and only a couple of times a minute at that.

After a time, I find myself in the immortal world, ready to meet my mentor. There is no floating this time, no high-altitude views, no cavorting through the clouds like a drunken eagle. Nor is there a rapid transit across oceans, archipelagos, forest tracts, jungles, or deserts. Instead, I simply find myself standing on a vast plain of nothing but salted sand, vultures circling overhead. The morning heat is crushing, the air is still, and the substrate leaves me nauseated and dizzy. The sweat at my temples evaporates before it reaches my cheeks. I cannot float and I cannot fly and the sand squeaks beneath the white cotton soles of my black Chinese slippers as I walk. Schlieren makes the horizon shimmer. A slow-moving herd of kudu tells me I'm back in Africa.

The heavy edges of my monk robes drag in the sand, the salt dusting the black fabric. I loosen the frog buttons but feel no cooler. I head for a distant waterhole. Minutes blend into hours. A Bateleur eagle prances at the shore, black-and-tan wings flapping, red beak slurping water. I'm obviously back in Africa, but I'm not exactly sure where. Jackals eye tiny springbok. Wildebeest raise a cloud of dust and push impala and zebras from the shore.

A giant Nile crocodile breaks the surface and snags a springbok in his jaws. In the little antelope's eyes, I see the terror, pain, desperation, and finally the resignation that is the fate of all the world's prey, whether they be turtles or tigers, songbirds or children. The crocodile performs its death roll. Delicate hooves point skyward, and strips of meat and skin appear in the water between flowers of blood. Far away, elephants trod along a line of green. I cross the salty plain toward them and eventually find rocks and trees. I check a flat rock for scorpions or snakes, then take a seat. I survey the brutal landscape.

"Greetings, Monk," says a breathy, sensual voice. "How do you like Namibia?"

I look around but see no turtle. I figure I will before long.

"Thanks," I answer. "I wasn't sure where I was. So far, it's a bit daunting."

"I smell that on you. Gazelle turds, too, if I'm to be honest."

"Hard to avoid while crossing the salt flat."

"I know. I brought you in that way because I wanted you to get a good feel for the place."

"Excuse me, but I actually don't see you."

"Right here," she says, extending her head and limbs. She's so well-camouflaged and minuscule, I wouldn't have been able to pick her out of the landscape if not for the sudden movement. She has a pointed beak and a flat, reddish-brown shell with light rings around the centers of her scutes. She is a Berger's Cape tortoise, locally known as *Nama padloper,* which means "road walker" in Afrikaans.

"You're so tiny," I say.

"Smallest of the tortoises, and better suited to this unforgiving land than you are, Monk."

"I'm sure that's true."

"I consume only what I need," she says, blinking against the bright sun. "Imagine how much better the world would be if your kind did the same. Instead, you indulge every wish and whim, consuming everything in sight, ingesting it in one way or another and excreting it back into the land."

"You're mostly right, I think. Although some of us don't do this. The Masai, for example. Do you know them?"

"I don't."

"Really? They're an African tribe. They move constantly in search of food and resources and own only what they can easily carry—a spear, a few cooking pots, maybe some medicinal herbs, their clothes, headgear for the women. Their lives are rich in relationships and experience and their connection with the natural world around them; material excess is not for them, nor are they obsessed with the false security tied to the idea of abundance."

"Namibian tribes like the Nama, the San, and the Caprivians used to live that way before white people showed up—in pure accordance with nature."

"Speaking of nature, come up here and sit by me. It's awkward talking down to an immortal."

"Will you protect me? I've always wanted to get up that high, but Namibia is full of turtle-eating eagles."

"Of course I'll protect you. The glare of the sun off my bald head alone will disorient any predatory bird."

"Ha," she says.

Gently, I lift her. "Wonderful view!" she cries in a voice as dry and tiny as a cricket breaking wind. "We turtles are conservative. We experience things deeply and completely, but most of the time, we live minimalist lives with modest desires, longevity our only ambition. We don't unnecessarily expose ourselves to peril. I suppose I'm a bit of a turtle rebel, though, because I simply don't see the point of a life lived in constant fear—"

"The immortal turtles I've met haven't been so fearful," I interrupt. "I think you have a war going on inside you because you chose to incorporate in such a small body."

"What a wise monk you are," she answers.

As she settles into the center of my palm, a spiral of energy develops between the acupuncture point called *Laogong* and the spot on her plastron that once tethered her in her egg. I may be learning from her, but she is feeding on me.

187

"I know what you're doing," I say with a smile.

"The trees are not so tall as I thought they were," she says, gazing up at a tableau of leaves and clouds and pretending not to hear me. "But they are still just lovely. Tell me, what do you suppose drives the human desire to own things?"

"The mistaken belief that ownership will fill the emptiness we feel inside."

"Why do you feel so empty?"

"We've lost our way," I say. "Given free rein, we imagine all the ways we could be better, then flagellate ourselves when we fall short. The space between who we are and who we think we should be creates a vacuum into which we fall. Falling, we grasp. Grasping, we shop. Shopping, we buy what we don't need, most often with money we don't have. The resulting debt makes us desperate. Desperate, we need instant gratification—sex, social media, more shopping."

"Who can have and be everything they can imagine?" the little immortal challenges me. "Where do humans even get the idea that they have to be exceptional or special, as if just being alive on this amazing planet isn't enough?"

The fervor with which she asks this question gets me to thinking of the long-suffering lives of turtles. I wonder whether—through copulating or laying eggs or basking in the sun or enjoying a bloom of delicious wildflowers or a sudden surfeit of snails—turtles also sometimes want to be more than they are or have more than they do.

"The yearning does drive some of us to be better versions of ourselves," I say. "But I take your point. Most often, our 'shoulds' come from people who want something from us. The agendas of our masters—not good masters like you—are always exploitative and self-serving. We've built a society upon the idea of worshipping and obeying them."

"And look where it's gotten you," the immortal says, confident in my protection and perambulating the rock. "The best version of yourself is the creative, organic one, not the one that submits to distraction, disappears into virtual worlds to have pretend sex or pretend wars or pretend races."

"One of your friends told me that everything we do is pretend, that it's all a game."

"I'm trying to say that a scripted life is not a fulfilling life," says the immortal. "Even if the script is written by a genius. We all have to be who we really are, not who others would have us be."

"Truer words were rarely spoken."

"Having said that, since you're not immortals, you have to pay attention to your own survival. If you're dead, you can't accomplish self-realization. Survival requires discipline, persistence, and effort. Every turtle knows this. We turtles pay attention to what we're doing. We own the consequences of our actions. Now, would you do me a small favor?"

"It would be my privilege and honor."

"Thank you. You're a courteous monk, I'll give you that. Would you mind going over to that bush of beautiful Nara melons and opening one for me? I love the taste, but I can't bite through them until they are soft, by which time they are rotten. I have always wanted to taste a fresh one."

I set her down atop the rock and make my way over to evaluate the fruits. The immature melons are about the size of a golf ball, green, round and knobby; the more mature ones are slightly larger but softer.

I pluck one. It smells exotic, aromatic, and tantalizing, with large dark seeds and custard-colored flesh. I am about to take a bite when I hear a small cry. I turn to see a medium-sized ground bird with a wickedly curved red beak, a Monteiro's hornbill, snatch up the tiny immortal.

"No!" I scream.

The hornbill lets go as I close in. I see that one of the turtle's forelimbs is gone and the stump is bleeding profusely. The hornbill snatches up the tortoise again. I take off in pursuit, my monk's slippers tapping quickly across the red dirt, but I cannot close the gap between us. Reaching a safe distance, the hornbill drops the tortoise again and plunges its beak into the little turtle's plastron. Intestines seep from the wound.

"Don't die! I'm coming!" I yell.

The little tortoise puts out her head. "Thank you so much for the melon. It was delicious!"

"No!"

"To live and die is better than not to live at all," she calls to me, her voice stronger than I expect. "Better this real world with all its mortal risks than the fake ones with all their dull safety. Seize the day, Monk! Be who you really are and be happy."

An instant later, as her gaze pierces me, the hornbill bites off her head.

I collapse onto the dirt. Clouds gather as witnesses. Lizards scamper, grateful their own turn was not today. In the distance, a rhinoceros roars, an elephant trumpets, and a zebra screams. The hornbill points its beak skyward again and swallows. Africa turns dim again, and I return to my park, my heart heavy and my cheeks wet.

"That lesson I got loud and clear," I say, but it's still early in the park, and there's nobody about to hear me.

QI IS VIBRATION AND VIBRATION IS EVERYTHING

Today, I meditate atop a hotel in Kyoto, Japan. I flew in last night to spend some time at the temples here, so I'm sleep-deprived and disoriented from jetlag and time confusion. I'm not supposed to be on this roof. It's not a guest area. I'm up here because I had an idea the sunrise would be worth watching, so I poked around a little bit and found a set of service stairs. The roof is crowded with equipment and, strangely, a little bit of AstroTurf where someone has made a little putting green for golf practice. I've brought along a traveling tai chi sword with me for practice and decide to employ it in moving meditation practice. Carefully, I establish the perimeter of the roof so that I don't tumble to the pavement far below in the thrall of an immortal trance.

There are gutters at the edges, actually, but I am not sure they would hold my weight if I grabbed at them. The thing is, coming in from the perimeter leads me to an obstacle course of heating and cooling equipment for the hotel, ventilation shafts and such, making for a poor course for walking with light attention. It takes me a little time to find the right distance from the center of the roof and the right distance from the edge, too. I pace off the course with eyes open, making a mental note of any potential stumbling blocks, conscious of the fact that it's a bit silly to take such a risk, but confident that even in an immortal trance, my feet will know the difference between solid ground and air.

After a few circuits, I pause to watch the sun rise over the surrounding green hills. I hold my sword the whole time, feeling a connection to the land and the history even though my own lineage is Chinese, not Japanese. Once the sun is high enough to feel its warming rays on my face, I begin my meditation, my eyes barely open, taking one slow, measured step after the other, allowing the sword to circle me from one side to another in a regular, hypnotic fashion that just deepens my trance.

And I'm there. And it's still dark. Little hands touch me everywhere—grasping, pulling, probing, squeezing. I feel as a warrior ant returning to the mound after battle, or Gulliver awaking in the land of the Lilliputians. Whatever grasps at me chitters and chatters softly. I wonder if I've landed on another planet and consider the possibility that turtles, rather than following one of the convoluted, much-contended paths suggested by evolutionary biologists, are in fact alien imports who arrived here by meteor, interdimensional travel, or as companions to some space-ranging magi.

That's when I detect the unmistakable aroma of vanilla. There is no chemical undertone to the fragrance, no alcohol to suggest it's vanilla in a seasoning bottle or vanilla as perfume. No, this is vanilla as it wafts delicately forth from the plant, a scent I recognize because I once visited a vanilla-orchid farm on the island of Huahine in the South Pacific. As I inhale deeply, the fragrance gradually becomes bitter and acrid, but familiar and agreeable. It takes me a few breaths to realize I am now sniffing coffee. After so much time in Asia, I've developed a real taste for fine tea, but the occasional cup of strong, fresh joe remains a guilty pleasure for me.

I smile at the silly machinations of my own mind, my little preferences and judgments, and as I do, foul-tasting little fingers take advantage of the opening and push their way into my mouth. I blink, the world grows light, and a dozen animals come into focus. They are of two groups: one with dark faces against white fur, the other with gray bodies and brownish heads with a vertical black stripe across the snout. All have long tails and monkey-like bodies. At my feet is a male tortoise perhaps thirty centimeters long, with a tan, highly domed shell, its neck stretched in my direction, its eyes focused on me.

"My guardians don't know you, Monk, although I've been expecting you," the immortal says, speaking very slowly, every word drawn out the way people who don't know turtles believe they all think and move. "It's their job to investigate and protect."

"Well, please tell them to take their fingers out of orifices."

He does so and his minions back off. When they do, I can see that I'm in the remains of a forest alongside a badly polluted bay.

"In case you are wondering, the big white ones are Decken's sifakas, the smaller ones are Western bamboo lemurs," the immortal tells me.

"Lemurs? Are we in Madagascar?"

"We are."

"Are you perhaps embodying an Angonoka—the rarest tortoise in the world?"

"I am."

"How marvelous! To actually visit such a legendary rarity and in such a beautiful place!"

"There used to be plenty of my kind, and this indeed used to be a beautiful place. One of the most splendid of all Earth's islands. A great paradise. No more. Tragically, this is a wasteland now, more of a heartbreak every day."

"Overpopulation," I venture. "Plundering of the wilds."

"As it is all over the world, but nowhere more than here, with our sapphires and our timber and so many other natural resources."

"It may sound strange, but as I materialized here, I thought I smelled vanilla, and then coffee."

"Madagascar is critical to the world's coffee supply," the immortal Angonoka replies. "Number three of the famous human vices after sex and alcohol."

"I didn't know there was coffee in Madagascar."

He laughs. There hasn't been all that much turtle laughing on these meditative journeys of mind, and the sound is most welcome. "Madagascar's coffee is critical to the health of the world coffee market. We have varieties here not found anywhere else, strains of the coffee plant that are resistant to parasites, pests, and botanical diseases. It's a repository for genetic variation unlike anywhere on Earth."

"A hedge against crop problems in the rest of the world," I say slowly. "Strains of the plant that can be crossbred to produce disease-resistant hybrids."

"Some of our coffee has a high caffeine content," the immortal explains, standing up on his limbs so that he looks like a little brown pyramid rising from the burned, dry remains of what was once a beautiful forest. "Some of it has almost no caffeine at all. Like our animal species, everything began on the African continent, moved here, and diversified in a beautiful biological explosion. The coffee plant diversified about 500,000 years ago. Human beings started mucking about here about two thousand years ago."

"Slash and burn, right? I remember reading about that somewhere."

"Natives call it *tavy*," the Angonoka replies. "It's like emptying your bowels into a ditch. If one person does it because he can't make it to the toilet, no big deal. If the whole city does it, you've got a problem. Instead of cutting sparingly and waiting for the forest to grow back, people here are constantly cutting everything down, utterly destroying the place."

"Have conservation movements not taken hold here? Seems to me there should be all kinds of NGOs and nature groups here by now."

The word "conservation" seems to inflame the lemurs who begin hooting and dancing. Their display goads the sifakas, who begin the most extraordinary vertical leaps.

"Look at them," I say in awe. "They're like booster rockets. I bet if you got them excited enough, they could reach right into outer space."

"NGOs are here. Conservation groups are here. None of them can match the persuasive power of logging profits. Industry just buys politicians, parties, elections, votes, whatever they need to do. And speaking of my guardians going into the galaxy, you do realize that your own presence here invalidates the so-called 'Fermi Paradox'?"

I shake my head to clear it. "The what?"

"You know about booster rockets but not about Enrico Fermi?"

"A physicist, right? Had crazy ideas about electricity?"

The sifakas and lemurs quiet down, and the Angonoka stiffly circumnavigates me, his nose to the ground as if he's looking for something buried under the topsoil. "That was Nikola Tesla, and his ideas weren't so crazy. Fermi built the first nuclear reactor. He worked on that marvelous contribution to life on Earth, the Manhattan Project."

"The atomic bomb. You're the first sarcastic immortal I've met."

"Then you haven't met many. Anyway, Fermi's paradox isn't about bombs; it's about alien visitors. He once famously asked 'where is everybody?' He figured that since there are countless stars, solar systems, and planets that could support life, including planets *older* than Earth, some alien civilization certainly should have visited us by now. The fact that they haven't is his paradox."

In the distance, a white Land Cruiser rumbles toward us, raising a plume of brown dust in the desiccated landscape. The sifakas and lemurs go absolutely crazy. "Here they come again," the Angonoka sighs. "I guess this is the end of our conversation."

"Here who comes? What do you mean?"

"The conservationists."

"They're coming for you?"

"Do you see any mates for me here? They've caught me seventeen times and brought me back to their compound. They want me to mate with a female they're holding, which might be a good idea since there are only a handful of us left in the world. Trouble is, she's a ghastly bitch. Truly horrible, with a voice like a shrew and a temperament to match. Self-righteous, judgmental, narcissistic, sure she knows it all. If you're not talking about her, you're not talking. Totally the wrong vibe for me."

"Wrong vibe," I repeat. "An immortal who talks like a 1960s hipster."

"The pinnacle of American culture, as I'm sure you know. Your Hippies had it right. Everything, and I do mean absolutely everything, really *is* about vibration. *Consciousness* is just vibration. Everything vibrates, see? Consciousness didn't just spring up in a turtle ancestor, mosquito or vole. It is omnipresent in all matter. Ubiquitous."

"We Daoists call it qi."

"Just so. And it's in atoms and dirt and rocks and trees just as surely as in porpoises. What makes some creatures possess more of it is the resonance that comes from complex, larger structures, and from the resonance those structures have with others' structures, living or not."

"Even if consciousness is an essential property of matter, there still must be some kind of difference between the consciousness of animate and inanimate things, yes? If not, then the very concept of something being alive is meaningless."

"You're correct," says the tortoise, moving away from the approaching Land Rover but too slowly to avoid capture. "In fact, the concept of life has great meaning. An aggregate of rudimentary consciousness, in a crystal mine for instance, remains rudimentary, while a biological organism resonates on so very many atomic and molecular levels that it reaches a far higher level of consciousness. Your scientists are learning to pick out particular proteins by their vibrations, and proteins are key building blocks of life. It's the interconnectedness within the creature, as well as its interconnectedness with the world around it, that determines its level."

"Even among people, I think there are different levels of vibration. Hold on, did you say you escaped seventeen times? Does that mean you came back here from a different part of the island?"

"Not here. I go someplace different each time. If I kept coming back here, they would just have to drive out and retrieve me each time. I'd have no solitude, Monk. I'd have no time to vibrate with what's left of the island. No, each time I go, I have the guardians take me someplace different. You have no idea how fast those sifakas can move, how much territory they can cover, how they swing from tree to tree and climb sheer rock faces holding onto me with one hand, using their tails for safety. I can cover miles of territory with them and we can go deeper into the remaining forest than these conservationists have ever been. I'm here this time because after trying so many other locations, it turns out this one is perfect. Nobody would think to look to me here, though the guardians are uncomfortable being exposed. There's so little cover here by the seaport. Even so, I've been here for months and would stay if someone hadn't randomly reported seeing me."

"I'm surprised they don't put a radio transmitter on you," I say.

"Oh, they do. And with stronger glue every time. But there is no glue too strong for a lemur to chew through. Once I'm free of it, we move. Then I rub my shell against rocks for days until all evidence of tooth marks is gone. I don't want them figuring me out."

The Land Cruiser pulls up in a cloud of dust. A beautiful blonde woman in safari top and shorts—more Venice Beach surfer than Malagasy—climbs out of the driver's seat. The guardians scream, leap on the vehicle, and harass the woman until she retreats into her vehicle. I see her make a phone call as she stares at us, at the turtle, at least, which is now surrounded by the guardians in a two-tier ring, lemurs on the outside, sifakas in close.

"I have an idea," I say, as the woman gets out of her vehicle again and aims a spray can at the primates.

"Now would be the time," says the immortal.

The stench of hospital disinfectant fills the air and the animals scatter in disgust. While she's doing that, I let the air out of her tires. She turns just in time to see them go totally flat. She gets on her phone again, making a disbelieving circuit as she talks. The lemurs take the opportunity to return for the Angonoka, gather it up, and scurry away. The woman throws up her hands in defeat as I follow the crowd off toward the jungle.

Refracted by the pollutants rising from factories along the bay, the light of the sun makes the air glow. I watch ships ply the port, eating away at this once-magnificent island with every journey, stripping it of every imaginable resource to feed the frenzied, consumptive appetites of the rest of the human world.

"Do we continue to vibrate after we die?" I ask the immortal.

"Immortals do," he answers.

"So that's the big Daoist quest, then. The preoccupation of all those classic texts. To continue to vibrate?"

"Such a simple lesson, isn't it?" the immortal answers, his eyes shining, his shell strong, his mind clear, his future certain, if uncharted.

"But wait!" I cry as Madagascar begins fading away. "How do I *do* it?"

"What do you think all your practice is for?"

Then I'm back on the roof again with only a sword and Kyoto for company. I sigh. They're not so bad, these two companions. Not so bad at all. I'll just have to make do.

And keep learning.

SPONTANEOUS, REDEMPTIVE EVOLUTION IS STILL POSSIBLE

I spend some time on writing retreats every year, not always to produce spirit-writing but always to think deeply about things and fill up some pages. Sometimes I think deeply about big ideas; sometimes I think deeply about small but very precise ideas; and sometimes, when the muses are with me and I'm very, very lucky, I think deeply about beautiful stories. The key ingredients in these retreats are solitude and having nothing to do but the work at hand. They're not social times. They're not teaching times. They're not times of monk service, and they're not tourist times either, although I usually conduct them far away from home, in foreign countries whose landscape and culture resonate with the work.

One thing I can't allow to intrude are politics or world events. The world constantly brims with weapons of mass distraction and sometimes—and writing retreats are exactly those times—I have to disconnect. That means disconnecting my laptop computer from the Internet so as to render it merely a glorified typewriter. It means that I choose places to stay that have no WiFi and no television, either. It means that should I go out to eat somewhere and find myself surrounded by people discussing a financial crisis, an environmental catastrophe, a political cataclysm, I promptly up and leave.

The world is awash in people who do not know how to pass the time doing nothing. It is filled with people who can't stand still. It is full of people who crave stimulation, even if that stimulation hurts or kills them. It is not, sadly, filled with meditators. Yet meditation is precisely the best antidote to what ails a world writhing in the pain of injustice and loneliness, poverty and war. It's also a great support system for a world filled with joy and adventure and love. It's a balancing force, one that deepens the good and counters the bad and I must do it even when I have created the physical space I need to write a book, because even when I'm alone, I've still got all that noise in my own head.

Today I stand in a mountain forest somewhere in Asia. Precisely where must remain a secret, because to share the detail would be to violate the sanctity of the retreat. In any case, there are some animals around, but they are shy because in this country anything that moves is hunted and killed and eaten, and anything that grows in one place is likely to be harvested too, one way or another, for food, shelter, decoration, or some psychotropic effect. Even before I ascend to the realm of the immortals for another Daoist lesson, I am aware of the animals and I am aware of the plants. How can I describe that awareness? It's energy and it has a kind of buzz. Not one I hear with my ears but one I detect with some different organ system, some different part of my brain, some different combination of senses. How am I sure it exists? Only by its absence. I don't feel it when I'm in an office building, for instance, nor in an airplane or a hospital.

There is rustling of leaves and the calling of birds and the croaking of frogs in this forest—all input separate and distinct from biological energy—and those things soothe me. I have to adjust my feet from time to time and shift the way I use my hips because the ground I'm standing on is uneven and if I'm to be as straight-spined as I like to be while meditating, I'm wont to tip over. All the same, because nowhere can we find a complete lack of distractions, I find my way into a trance, and into an open field, in mud up to my ankles, a line of beautiful, bark-mottled alder trees standing sentinel, the familiar smell of wetlands in my nostrils.

The sun is bright and the air is warm. A light breeze penetrates the perforations of my jade-studded monk's hat. A cacophony of flies seems bent on laying eggs in my ears, and it is only after repeatedly brushing them away that I realize they are buzzing to the tune of Troy Caldwell's "Heard It in a Love Song," made popular by the Marshall Tucker Band. I put this together with the sphagnum and cattails and conclude I'm in the American South. When the flies add another '70s offering, the Charlie Daniels Band belting out "The Devil Went Down to Georgia," I get the message loud and clear, though it isn't midnight and I didn't arrive here by train.

A rare indigo snake slithers past, two meters long and resplendent in iridescent black scales. Songbirds bicker. Thunder booms in the distance, but I see no clouds. I feel the urge to trudge along and look for my turtle mentor, but the depth of the mud discourages me from taking even a single step. The flies change to backwoods bluegrass now, insect banjos loud in my ear. The sun transits slowly, directly overhead now, beating down with a fervor. Beads of sweat caress my temples on the way down to my white collar.

"I'm here!" I cry at last, figuring it can't hurt to announce myself.

"And so am I," a small voice replies. "Been waiting on you here for a while, but I understand that of late you have had other important places to be."

I look down and spy him at once, a diminutive creature with a round, rugose, reddish-brown shell, a lighter colored center to each scute. It is a male North American bog turtle, the rarest turtle species on the continent, an object of legend to conservationists and the greatest possible lust to domestic fanciers.

"I thought all of you were gone," I say. "I remember way back when I was an intern at the Bronx Zoo, my boss, the curator of reptiles,

SPONTANEOUS, REDEMPTIVE EVOLUTION IS STILL POSSIBLE

said bog turtles would be extinct within ten years if we didn't do something."

"Well, he did something," the immortal replies, wiping an eye with the soft side of a leg. "Lots of others did, too. Preserves and laws, regulations and enforcements. Rules and more rules, which only the law-abiding care about. Even so, we've got our secret ways, don't you know? Always have, always will, at least until the end of days, which, of course, is coming soon."

"Ah," I say. "Am I here to learn about the impending apocalypse?"

"Let's take it slow," the little turtle says when I crouch down to it. "Follow the trajectory of the folly of man, if you will."

"I'm a little tired of doom and gloom," I say. "What about a good joke?"

"A joke?"

"Humor is a Daoist tradition, not that you'd know it with what I've been learning from all of you lately."

"Why did the dead baby cross the road?" the immortal asks.

"Dead baby jokes are in terrible taste," I say. "Not funny. Really."

"Because he was stapled to the back of the chicken."

I have to turn away and bite my hand not to laugh. Really. I just don't want to. It's just wrong. But I can't help myself. I double over in my attempt to restrain a guffaw.

"Don't talk to me about man's folly," I say. "Talk to me about enthusiasm and love."

"You're not really here to dictate the lesson," the bog immortal says sourly.

"Just frame it for me in a more positive way, will you? I have to put the spirit-writing together in a way that won't drown readers in negativity."

The little turtle thinks for a minute. "I don't think humans are essentially negative," he says at last. "I think what happened is that happy-go-lucky people became too serious after the dawn of agriculture. In the wandering stage, before planting anything, there was time for love and laughing. Fighting, of course, too, but the fundamental need to protect and preserve property was missing, leaving emotional space for experience and bonding."

"Free and easy wandering is exactly how we Daoists describe the ideal life," I say. "We crave it, though it's not so easy to find anymore, nor to craft."

I hear thunder again but see no clouds. It's so odd. I want to ask the immortal about it.

"Sex," she says, before I have a chance.

"What about it?"

"There are young and adventurous humans, but most of you settle for too little of it in too predictable a pattern. Bonobo apes do better. Look at them for inspiration. All that creativity ended with the dawn of agriculture."

"On the other hand, intellectual freedoms blossomed when we had enough food," I say.

"For a while. Then all that agriculture became a burden. Plus, it sent human population out of control. It's time for you to stop trying to command everything and return to surrendering."

"Do you mind if we find some shade?" I ask. "These robes are so heavy in the sun."

SPONTANEOUS, REDEMPTIVE EVOLUTION IS STILL POSSIBLE

The turtle nods, so I pick him up and walk to the edging alders.

"Tell me more about surrendering," I say.

"It meant it in the sense of settling for less of some things while gaining more of others. Your ancestors accepted a poorer diet to be assured of regular, if not monotonous, meals. More precisely, they stopped chasing resources when they began planting crops, thus the complex and nuanced set of foodstuffs required to keep them in optimum health were no longer on the menu."

"I eat a lot of veggies," I say. "I'm like a hunter-gatherer monk, vegan and all, chasing seeds, greens, fruit, and nuts."

"Really? Most people I've seen who eat that way don't have such a belly."

"Nice," I say. "You know, I've seen fat turtles, too. You're not so much better than me. I see rolls of fat around your axillae, where your limbs go into your shell."

"Pure muscle," the bog turtle says, a bit defensively. "Turtles only get fat in captivity when we are fed processed junk instead of the food nature holds for us. It's like what happened to you people when you started caring more about your crops than you did about your bodies. The narrow task of sowing and reaping replaced walking long distances, hunting, gathering, and climbing, all the activities to which you were adapted, and which kept you happy and healthy and fit."

"The thing is, turtles don't have a choice," I say, settling in against a tree, the small of my back itching a bit because of the rough bark. "You have to endure because that's your architecture, which doesn't exactly entitle you to a position of moral superiority. Humans moved because we *could*. We planted because we *could*. I believe that our biological reality has a lot to do with our destiny."

"Being able to do a thing doesn't always make it a good idea. A bobcat can eat her young. A bee can fly straight into the sun and burn up in an instant. Possible, but not advisable. Wisdom is as much about knowing what *not* to do as knowing what *to* do. Too, being built for a thing doesn't consign you to do it forever."

"Well there we agree," I say. "I'm looking for some kind of spontaneous and transcendent piece of evolution so humans can survive in a beautiful world."

"That's what's needed," the bog immortal replies. "Anyway, your possibilities narrowed once your activities did. The gene pool people could dip into grew smaller. Shallower, too. Why? Because you just hung around. You didn't wander long distances as you were built to do, spreading your seed and the like."

"You're only talking about seeds because I did," I say, spying a berry on a bush, reaching up for it, and offering it to the little turtle.

"No, seriously," he says, his beak smeared blue as he consumes my treat. "With agriculture, your life became localized and you didn't meet strangers. You narrowed your gene pool and became inbred, something we turtles do with fewer ill effects than your kind does, though we get the occasional two-headed hatchling and such when we're stuck inside a perimeter by real estate developers or when a pond shrinks from climate change. You people got sicker, too, with rare or new diseases, all because your diet was deficient, and you didn't use your bodies they way nature intended. By the way, when I say disease, I don't only mean physical afflictions. Your brains got sick. You acted strangely. You become obsessed with your land, your stuff, security, not having to worry about food even though you substituted that with all kinds of other worries, like social status and protecting what was yours from other people who would take it from you. Once you started trading what you had with others, you developed class distinctions. You institutionalized slavery. You created your own hell in the name of not having to worry. This new torture you inflicted upon yourselves extended even to your unborn children, as the destiny of an individual member of society predetermined in an unimaginable way. What perversity."

"I asked you for a more positive take on humans," I say. "I'm not really feeling that here."

"This is nothing, monk. I'm just getting started diagramming all your wrong turns. Do you know what came next? After agriculture?"

"Industrialization?" I suggest timidly.

"Just so. That lovely phase in people history in which a privileged few enjoyed new heights of power, luxury, and privilege, while the exploited majority lived short lives of heavy labor in factories with toxic air and other crushing conditions."

"There's that thunder again," I say, shielding my eyes to scan the sky. "So strange. Not a single cloud."

"That's not thunder; that's a helicopter. They do the surveys from up there so they don't get their boots wet, then they come down and snatch us."

"What surveys? Who snatches you?"

"Scientists. They're obsessed with numbers and with morphological relationships. They get theories in their heads and hunt us to prove them. Get a grant. Write a paper. Get a job. Get tenure. It's all about them. They care so little for us that they pickle us in foul smelling liquid. I've seen jars and jars of my kind floating, dead and bloated, their eyes gone white, their limbs shedding wisps of skin, just so they can measure us, count our scales, cut us open to see what's inside us. Inquisition, that's the word. To a turtle, it just means genocide as it has to so many other species. Only monsters murder living things simply to understand them better."

"You're saying biologists are flying in to kill you?"

"If they find me, I die, so I'd better get back to your lesson, yes? Wouldn't want to leave before you clearly understand where your species has been and where it's going. So then, once you people industrialized, vast numbers of you began living and working in close proximity to one another. Pandemics could and did spread rapidly, taking millions of lives. All of us nonhuman beings became your victims on a new scale as you began to so-called 'terraform' our planet. You cleared forests. You dammed and diverted rivers. You restructured ecosystems according to your own pleasure and agendas. You synthesized religious and mythological justifications for murdering other sentient beings. The lessons of nature were lost on you, and with them all balance and harmony, and the loss of so much began in great earnest."

The thunder becomes a whump-whump as the helicopter comes into view. It's dark blue and bears an insignia in the shape of the State of Georgia, replete with a whited-out image of a deer, a fish, and a bird. It grows closer, lower, circling in apparent search of a clearing where the ground is firm enough to support it and the trees give quarter.

"See that sign? Department of Natural Resources. That's what deserts, jungles, oceans, forests, grasslands, plains, marshes, swamps, tundra, and taigas have become for your kind. Anything and everything that lives and grows is nothing more than a *thing* to be plundered."

"I don't feel that way."

"So what? You're just one little monk. Honestly, only the people close to you care how you feel. Your kind has chopped down, blown up, burned, flattened, stripped, razed, and utterly decimated the entire world, not just at the surface but below it. You people are a cancer, monk, and the victim is Planet Earth."

"I hope you're wrong."

"It's no longer a debate or even a discussion. It's the way it is."

The helicopter sinks below the tree line. I know it has touched down because I hear the engines shut down, first a slowing of the rotors, then a whine, then quiet. A moment later, two men appear at the edge of the clearing, not a hundred feet from us. One wears a gray shirt and green trousers and has a badge and gun on his belt. The other wears bush pants, hiking boots, and a tie-dye T-shirt. His hair goes past his shoulders, and his lanky frame and beard make him look like a Gen-Y messiah.

"Why do I see police?" I ask.

"If all the world is a resource and resources need to be protected, is not a worldwide police state the inevitable next step for your kind?"

"You're so harsh. I wish you could judge us with more compassion."

"What compassion have you shown turtles? Here comes the scientist, obsessed with classifying rather than appreciating, more interested in building mind castles than preserves, indulging intellectual curiosity instead of acting as protector and restorer."

"People do have their vanities."

Pudgy and with a sweat-stained shirt, the lawman leans against a tree while the biologist works the wetland, turning over logs, scanning hillocks, and crouching to poke puddles with a snake stick, a long metal rod with a grasping hook at the end. I hear a rattlesnake sound off.

"Hope it bites them both," the bog turtle says bitterly.

The police radio crackles. It's the helicopter pilot reporting that the chopper is sinking into the mud. The expression on the officer's face says he's glad for the excuse to cut his babysitting short. He tells the biologist they have to go, in response to which the young man plunges his hand into the grass and comes up grinning jubilantly, a five-foot rattler dangling and twisting angrily in his grasp. He drops the snake into a canvas bag and knots it tightly.

"One less snake in the bog," says the immortal. "Soon there will be none. Know where it's going? To some museum where it will 'contribute' to a scale-count study or some such nonsense. Then on to a giant metal shelving unit where it will stay until some clumsy rube knocks it over, mops it up, and throws it into a dumpster."

"Let's make sure you're not next," I say.

"Nothing you can do about it. That kid's good. He's been here before. He doesn't miss a trick. Eyes like an eagle, persistent as a panther. If he's this close to me and I'm out of the mud, I'm done for."

I try to step on the turtle, gently, to push him down into the mud. I'm too late. The biologist sees the turtle and strides over, his speed hampered only by the pull of the muck.

"This is the end of me," the turtle says. "Before I go, use your spirit-writing to catalyze the transcendent evolutionary change you mentioned. Choose a new version of humankind rather than the thundering oblivion of global climactic catastrophe and atomic Armageddon. Learn the difference between destiny and fate."

"Come on," shouts the lawman. "Our ride is sinking! Chopper pilot's nervous. We gotta go!"

The young scientist raises his patience finger. He's upon us now, and though I attempt to shield the bog turtle, he moves right through me, an uncomfortable feeling to say the least. He feels something, too. His face snaps shut like a clam. He pauses, shrugs, and picks up the turtle.

"Talk to him!" I cry.

"He can't hear *me* any more than he can *you*," the turtle answers faintly, his head inside his shell. "My kind has been imploring your kind for mercy for millennia beyond imagination. It's no use. It has never been any use."

"Please try," I implore him as the biologist turns the little turtle around in his hand, tries to pull on a leg, turns him on his back, runs his fingers along his marginal scutes.

"STOP!" the little turtle screams.

The biologist suddenly looks as if he's stuck his tongue in an electric socket. His long hair flies up and out, and his T-shirt, advertising a rock band named for an aardvark, floats away from his skin. The bog turtle tumbles from his hand, landing face first in the mud, tail sticking up in the air.

"What the..." the biologist whispers.

"Keep talking to him!" I say, "Make sure he knows your voice is real."

"I don't want to be pickled!" the bog turtle yells, righting himself. "I want to live out my life where I am, not be a victim of your curiosity. I'm not a toy. I'm not a rock. I have a family. I have ancestors and memories and I take joy in the crunch of a frog's leg between my jaws and in sinking my beak into the bulb of a water hyacinth or lily. I want to seduce my mate and consummate the deed. I want to see my children hatch and grow. I have a right to all that, and it is as real and strong as your right to do the same things in your life and far more real than your right to kill me. Now go away and leave me alone!"

The biologist collapses to the ground and sits there staring, his pants stained, his boots dripping, the bag containing the rattlesnake flopping one way and then another as muddy water seeps in. He runs his hands over his face and over his forehead, pushing his locks out of the way and painting his face with mud like a Native American warrior.

"This can't be," he mutters. "It's my own subconscious. My guilt."

"It's the turtle!" I bellow, even though he can't hear me at all. "It's not about you!"

The bog turtle sinks into the mud. The lawman tromps over and grabs the scientist by the elbow.

"Don't ignore me! We're outta here right now. Come back on your own if you want."

"The turtle spoke to me."

The lawman yanks the younger man to his feet. "I don't appreciate babysitting potheads," he says. "I got real work to do."

I feel badly for the biologist, the way he keeps looking back over his shoulder at the bog turtle as he's dragged away.

I bend to the turtle. "You see?" I say. "Miracles are possible. Spiritual evolution is possible. If the history of Planet Earth is about anything, it's about unpredictable change."

The look the immortal gives me says that the lesson went the other way this time, and that I've succeeded in at least shaking his impossibly pessimistic view of humankind.

Then I'm back in the forest, meditating, the wind blowing, the energy of all those living things buzzing around me.

Inexplicably, I find myself filled with irrepressible joy.

MEET THE TURTLES

You've already met the immortals, the sage mentors who chose
to inhabit reptile shells in order to appear, at least to me, in a
meaningful, familiar, and accessible way. Now it's time to meet the
turtles themselves, the real-world animals who have served as my
literary foil in these many pages. Of course, if you are taking this
book in the manner in which it is most seriously intended—as spirit-
writing—you may choose to skip the natural histories outlined below.
If you do so, I won't hold it against you, but don't you think that after
having done such a yeoman's job of shouldering the burden of tens of
thousands of didact words, the real animals deserve a little attention,
too? Don't you think it's worth knowing about their very real and
mortal challenges in the world from a scientific, or at least a nature-
lover's, point of view? I hope that you do, and that you will grace the
details of their plight with your attention. After all, what better place
to put all the lessons of this book to work than by treating all suffering
and dying creatures with compassion?

Aldabra Island Giant Tortoise

If there is a single species that most embodies my lifelong love of
turtles, it is the Aldabra Island giant tortoise, *Dipsochelys elephantina*.
Readers who recognize the Latin prefix *dipso* for its association with
alcoholism will be happy to know that these spectacular turtles are not
drunkards, but rather are able to drink water through their nostrils,

an evolutionary feature essential to their survival on hot, arid Indian Ocean coral atolls, such as the nominate island of Aldabra and others in the Seychelle Archipelago. Along with Galapagos Island tortoises, Aldabras are an example of island giantism, a fancy way of describing the evolutionary move in the direction of great size when predators are limited and resources abundant. Although smaller than the giant ocean-going leatherback, these are the largest terrestrial turtles in the world, reaching more than three hundred kilos. A large male Aldabra tortoise is a breathtaking beast, with limbs like tree trunks and a neck and head that would not be out of place in *Jurassic Park*.

I don't love these tortoises solely for their great size, although it is truly impressive to watch a herd of them emerge together from the shelter of shade as daylight begins to wane. They graze, fall asleep, awaken at dawn to mate and relieve themselves, and then trundle back to the protection of trees. I am actually most drawn to them because of the many decades I have spent caring for them in my backyard, raising them from the size of tennis balls to large enough to stroll nonchalantly through my palm garden with my son riding them and scratching their heads. I have seen them through the occasional illness, cleaned the messes they made, and repaired the fences they ruptured. I have protected them from freak freezes and howling hurricanes, fended off racoons and rats and feral felines to keep them safe, and brought in (and paid for) truckloads of vegetables, cacti, fruits, and forage for their dining pleasure. I have sat on the ground surrounded by them, watched them for countless hours, and I have come to love their calm, equable personalities, the gentle way in which they show affection, their amusing territorial displays, their distinct personalities, preferences, and habits.

The Aldabra Island giant tortoise to whom I am most connected I call Nobuko. My others also have Japanese names, all for personal reasons having nothing much to do with my life as a monk and nothing much to do with the tortoises either. I am well into my fourth decade with

Nobuko as my companion, and I believe it is fair to say I know every inch of her and every one of her expressions. Perhaps she says the same of me to all newcomers to my yard. One year, she broke through a fence and ended up in my modest swimming pool. It took quite a few helping hands to lift her back out, although she was perfectly safe and happy in the water. In the wild, Aldabra tortoises are actually excellent swimmers and can navigate the ocean between islands in the Seychelle Archipelago without much trouble. Presumably, this is the way their ancestors arrived in the islands from their original African home.

One source in my turtle library claims that Aldabra Island giant tortoises range from bronze to black in color, but I've never seen one that was not either dark gray or black. They have narrower heads than comparably sized Galapagos Island tortoises, and, at least to my eye, more expressive faces. Their skin tones match their shell color, and both these match the color of the coral underpinning their island home. They lay eggs in late summer and early fall, and many of their tennis ball-sized hatchlings are eaten by rats and crabs, while others don't survive the challenge of the dry heat. They populate a few different Seychelle Islands and can live one hundred years or so. There are perhaps 100,000 of these tortoises left on the nominate island of Aldabra, making their biomass per square kilometer the highest of any grazing animal on Planet Earth.

It may sound as if these island giants are thriving, and compared to some of the other species chronicled in this book, they are. All the same, they are in demand for pets, farmed and exported from the islands, despite being listed on Appendix 2 of CITES (a sign their population levels are of concern to the folks running the Convention for Trade in Endangered Species) and despite, too, their high price and very particular husbandry demands. What concerns me most is that the great preponderance of these animals live on a coral atoll that faces the existential threat of a surrounding, rising ocean.

Alligator Snapping Turtle

Plenty of turtles will snap. Mud and musk turtles, for example, are notoriously nippy. Softshell turtles, particularly the largest species, are predatory hunters that will not hesitate to defend themselves when molested, and both the Chinese bigheaded turtle, *Platysternon megacephalum,* and diminutive stinkpot, *Sternotherus odoratus,* will go for a finger when freshly snagged from a stream. Yet because they have lost so much of their protective shells over the long throw of evolution—both plastron and carapace are greatly reduced and soft parts are vulnerable—snapping turtles employ extra-powerful jaws, a sharp, penetrating beak, and a ferocious attitude to take biting to a whole new level.

Snapping turtles proper belong to the genera *Chelydra* and *Macrochelys,* and are citizens of the Americas, reaching as far south as Ecuador and as far north as Canada. Striding along the bottom of murky watercourses by day and by night, they hunt everything from crustaceans and fish to waterfowl, and even alligators! They are substantial turtles, and their presence on roadways is always impressive. The alligator snapper, *Macrochelys temminckii,* is the largest of the snappers, and the sole representative of its genus. Living in rivers that drain into the Gulf of Mexico, in Texas, and in some southeastern states, it is the largest freshwater turtle in North America. How large? A famous giant allegedly reached 227 kilos.

The shells of these holdovers from an antediluvian age render them all but invisible, particularly in murky water. Algae can grow on those shells, and barnacles, too, when the water is brackish. Leaving the water only to nest, alligator snappers spend most of their lives buried in the mud at the bottom of streams, ponds, rivers, and lakes. Walking the bottom less than other snappers, they hunt using concealment and surprise, opening their prodigious maws and wiggling an appendage at the tip of its tongue that looks like a pink, juicy, wriggling worm. When a fish darts in for a meal, it instead becomes one. There are other sneaky hunters in the turtle world, but the alligator snapper takes the prize.

During the last half century, and before it became illegal to do so, I kept and raised a few of these amazing giants, returning them to their natural habitat when they outgrew my ability to care for them. I always found that they treated me with the same respect (if not affection) with which I treated them. I fed hatchlings worms and minnows, and larger fish, beef heart, and shrimp as they grew, eventually offering rats. When I was a zoo intern in the 1970s, there was a large alligator snapper in the collection. As of this writing, the animal is still there, all these years later, bigger and badder than ever, ready to snap a broomstick or remove a set of fingers. There are rumored to be other giant old alligator snappers living in the bayous of Louisiana and watercourses in Texas. While they are now protected by law (as of 2019 the IUCN considered considers them vulnerable) poaching for Cajun restaurants and collection for turtle fanciers worldwide continue to pose a great threat to the survival of the species.

Angonoka Tortoise

Years ago, I visited with a few the Angonoka, or ploughshare tortoise, *Astrochelys yniphora*, at an American facility belonging to the Wildlife Conservation Society, the international preservation, education, and propagation organization birthed from New York City's Bronx Zoo. Seeing one of the world's rarest tortoise up close was an indescribable thrill for a lifelong turtle lover. This imposing animal is the same maximum length as the marginated tortoise, but heavier, at up to eight kilos. The extra weight comes from its highly domed shell, a shape it shares with its close cousin, *Astrochelys radiata*, the Radiated tortoise, long a darling of the pet trade in Europe, Japan, and the USA, though also highly protected by law. The Radiated is a smaller, more colorful tortoise, with a striking pattern of yellow radiating lines on a black background. The Angonoka, by contrast, starts out with some contrast between its generally tan carapace and rectangular black markings, but most of the color fades with age until it matches the local brush where this shy species spends its time hiding, emerging to forage and mate at dawn and at dusk.

I have not spent much time in the company of this species. Precious few Western fanciers have. If its radiated cousin is any measure, though, it is a reclusive animal not at all like the marginated tortoise or any of the variety of amphibious Asian species of which I am so very fond. Think of it as a beauty queen with a retiring personality, and one whose future is very much in doubt as a consequence of deforestation, introduced predators, and agricultural wildfires. Malagasy people do not eat them, and even superstitiously regard them as helpful agents in warding off disease.

Anganokas live in bamboo scrub, in a hot, tropical environment that isn't exactly dry and isn't exactly wet. Of personal interest to me is that many of the plants and animals that naturally attract me, including so-called day geckos of the genus *Phelsuma* and my much beloved Bismarck palms, come from Madagascar and are elements of the Angonoka's natural world. The tortoise's strong beak suggests it might tackle some of these bush plants as food, but they are also known to consume eggs and carrion.

The domed shell of an Angonoka tortoise may well have evolved as a defense against predators, since it is harder to bite and crack or penetrate a domed shell than a flat one. That dome, however, has some liabilities. Not only does it mean that the Angonoka takes longer to warm up than a similarly sized flatter animal, it also means that, should the tortoise flip over, its legs won't reach the ground to help it flip back over. If it can't use its head to right itself—if the ground is soft or there is nothing to push against—the animal can be stuck on its back until it dies, either from exposure, hunger, or thirst.

A word about the humidity relations of small turtles is probably in order. Small animals lose water at a higher rate than large ones, so baby turtles invariably stay either in water or in the most humid possible niche until they grow. This is because small animals have a higher surface to volume ration than large ones and therefore heat up and cool down faster. This principle is what helped my Yale mentor, Dr. John Ostrom, famously deduce that dinosaurs could not depend upon the sun to warm up as they would take so long to do so, night would fall while they waited to be warm enough to move. This deduction led to a revolution in the way we see dinosaurs, which are now thought to be closer to birds in their physiology than to reptiles.

Like the marginated tortoise, male Angonokas have a large gular scute—think of it as a lever projecting forward under its chin—it uses to overturn the female in mating, and perhaps in ritual combat with other males as some other tortoises do. Females may lay four or five clutches, each a month apart. When it rains, they lay one to six large, spherical eggs that take eight months or so to incubate. They live between fifty and one hundred years and have been the subject of conservation efforts not only by WCS, but also by a British group, the Durrell Wildlife Conservation Trust, on the Island of Jersey. This latter conservation group was founded by writer and conservationist Gerald Durrell. His colorful life with animals on the island of Corfu is rendered in a series of popular books, and there is a recent TV show profiling his family. The IUCN, as of 2008, lists the Angonoka as critically endangered.

Asian Narrow-Headed Giant Softshell Turtle

Though many turtles feature interesting specializations, arguably the strangest looking turtle in this book is the present species, *Chitra indica*. The Asian narrow-headed giant softshell turtle has a strong, thick neck terminating in a strangely tiny head. Its shell lacks the hard scutes other turtles possess, instead being comprised of underlying dermal bone covered by soft, leathery skin. The plastron is creamy pink and the carapace is olive green with light reticulations that extend up onto the head. These latter markings make the turtle appear as if someone has drawn lines on it with a pen, and may account for the animal's scientific name, as the word *chitra* means "image" or "portrait" in Hindi. *Chitra indica* ranges from Pakistan, India, Bangladesh, and Nepal.

Like alligator snappers, softshell turtles are sneaky hunters that remain buried in river sand until unsuspecting prey swims past. They lack the snapper's tongue lure, though, and unlike the snapper are fast, powerful, efficient swimmers fully capable of chasing and taking down prey. The softshell's war with the fisherman in my fable is not really so fanciful, for the Asian narrow-headed giant softshell is reported to grow (albeit rarely) as long as two meters, and at such a size would certainly be capable of capsizing a coracle. Softshell turtles are flattened dorsoventrally (top to bottom) so they don't encounter much resistance when moving through the water. They are fast, agile, and able to change direction easily, making them formidable predators.

Unlike Asian box turtles, *Chitra indica* are as unhappy and ungainly ashore as they are graceful in moving through the turbid, fast-running rivers they call home. They are so committedly aquatic that they can respire through specialized skin while at rest, extracting oxygen from water like a fish. As an aside, some turtles indigenous to the American Midwest and Northeast also breathe this way, sometimes all winter long, allowing them to survive in iced-over ponds. In this situation, the heartrate and overall metabolism is reduced to levels so low, it's hard to detect life in the animal, although others of those turtles have actually been seen moving slowly about on the bottom of such frigid bodies of water, suggesting that this ability is well developed enough to allow them to do more than sleep until spring.

The Asian narrow-headed giant softshell turtles are diurnal and prefer live prey, but when they're hungry, these turtles can be opportunistic, sometimes eating fallen plant material. Besides fish, they favor mollusks and crustaceans. This species grows too large to be kept in captivity, where, as a baby, it requires soft sand for burrowing and scrupulously clean water. The challenge to wild populations of the Asian narrow-headed giant softshell turtle is not that turtle fanciers over the world wish to fawn over them as prestige pets, but rather that people want so badly to *eat* them. Their flesh is reputedly tender and favored in traditional recipes. Sadly, this means the turtles are widely available in many Asian markets. They also happen to frequent rivers assaulted by industrial pollution, sometimes on a massive scale. With so many pressures upon their populations, they have become one of the turtle species most under siege anywhere in the world. The IUCN, as of 2019, lists them as critically endangered.

Berger's Cape Tortoise

There is something special, delicate, and precious about the little tortoises of the genus *Chersobius*. It's tempting to see this "something" as merely an expression of their diminutive size—frankly, they are adorable—but I see their appeal as something different. In them, I see the tenacity and strength that makes turtles what they are, the unique combination of anatomy, physiology, and behavior that has pulled them through countless challenges over hundreds of millions of years. One can say of an animal like the Galapagos tortoise, for example, that it has survived and diverged through an accident of isolation, lack of predation, and achieving giantism. One can say about the leatherback turtle, another impressive beast, that it has been refined by the very specific demands of the marine environment and evolved all the same physiological characteristics and hydrodynamic lines as many other denizens of the deep. When it comes to such a tiny creature as the Berger's Cape tortoise (*Chersobius bergeri*) and its kin, however, what we can say is that a life lived quietly, stoically, secretly, and without engaging extremes or indulging evolutionary adventurism has paid off. These little tortoises are, quite simple, the quintessence of turtle.

The genus *Chersobius* is a relatively recent moniker. Previously, the genus name was *Homopus*. It was as *Homopus* that I first met them, back in the 1970s, when the zoo at which I worked received a pair of *H. boulengeri*, a related and similarly delicate little tortoise from South Africa known as the Karoo dwarf tortoise and one exceedingly rare to see in captivity. A healthy turtle (any turtle at all) should feel as a rock in the hand. Any sense of hollowness or lightness is, to the experienced turtle keeper, a sign of trouble or worse. When the zoo's five *H. boulengeri* arrived, they felt like grayish-blue potatoes, but skin only, half the flesh gone. Their shells were not yet soft—a sign of severe malnutrition, disease, the leaching of minerals, and of impending demise in a turtle—but they were alarmingly weightless. Believing them to be dehydrated, we put them in a tropical section of the zoo, with high temperature and high humidity. Two of the five died within a week.

Feeling there was nothing left to lose, I prevailed upon my boss to let me try a different husbandry approach. I set them up in a dry box with rocks to hide, and a dim, infrared heat lamp that offered high temperature without any humidity. I gave them hiding spaces set up with rocks from my yard, disinfected first, and a substrate of cypress mulch. I offered no water bowl at all, left them in the quietest spot in my house, spent a small fortune on as many exotic greens as were available back before gourmet markets were born, and surreptitiously snipped and pinched off a variety of buds from bouquets at the local florist to tempt them to eat. I simulated morning dew with a mister, too. Within a few days, they were eating, and within a few weeks, I had *H. boulengeri* eggs.

Berger's Cape tortoise has not only a received a new genus name of late, but its species name seems to alternate between *bergeri* and *solus,* depending upon the scientific source. It seems to be the rarest of these little-known tortoises, and either the only one, or one of two species, that occurs in Namibia. Its natural history is poorly known, in part because, like the Tanzanian pancake tortoise whose flattened shape it resembles, it spends lots of time wedged into rock crevices and therefore out of sight. As in my fable, it often falls prey to birds, though hyenas eat it as well. Berger's Cape tortoise is thought to be mostly vegetarian, but like so many other turtles, particularly those living in arid, nutrient-poor areas, it is likely an opportunist and thus unlikely to pass up dead insects or carrion. People who try to keep it alive in captivity say it does not do well if deprived of the native plants and flowers upon which it feeds in the wild. The IUCN, as of 2019, categorizes *Chersobius solus* as vulnerable.

Central Asian Tortoise

Agrionemys horsfieldii occurs as far west as Iran and as far east as China, as far north as Russia, and as far south as Pakistan and the Gulf of Oman. Captured by the hundreds of thousands across its uncommonly vast range and shipped globally, it is a tragically and criminally abused species. In Europe, it is marketed as a garden tortoise, a euphemism for an animal that requires no care and can simply be left outside to die. In China, it is a food item. In the USA, Central Asian tortoise can be found stacked ten-deep in reptile wholesaler bins, doomed by an improper diet and inadequate care in chain pet shops. While it still exists in large numbers in the wild, the history of human/animal interaction teaches us that plentiful species quickly become rare, a fate sure to soon befall this quintessential tortoise.

Central Asian tortoises are charming creatures, small, oval, and spry with bright eyes and a gregarious manner. They are also enthusiastic diggers with bendable limbs well suited to maneuvering in tight spaces. They survive the brutally long, frigid winters in their range by excavating extensive, deep underground burrows. There is some speculation that the fluids in their bodies contain molecules that lower the animal's freezing point. While that notion certainly stands to reason, I haven't been able to find research citations about this. At one point in my early career as a student of zoology at Yale University, I translated a fair number of herpetological papers from Russian to English. Russian researchers painted this species—also known as the Russian, Horsfield's, and Afghan tortoise—as stoic creatures that bear tremendous seasonal hardships, emerging from their winter quarters to a springtime festival of feeding on wildflowers and grasses, elaborate courtship rituals (dancing, bobbing, biting, circling), mating, repeated egg-laying, and then a return to their dens for eight more months of half-frozen sleep. I can only hope the warm months bring a genuine orgy of pleasure for these tough little tortoises.

And the former Soviet Union really *did* send one of these turtles into space...

Fly River Turtle

I met my first Fly River turtle working at the Bronx Zoo, the same place I met my first alligator snapper. These turtles are both standouts in terms of behavior and appearance, evolutionary adaptations, and the way scientists choose to classify them. *Carettochelys insculpta* occurs in northern Australia, in Arnhem (aboriginal) land, and in Southern Indonesia and Papua New Guinea. They prefer fresh water but can tolerate a bit of brackishness, sometimes venturing close to the ocean in estuaries and river deltas. Females, which grow larger than males, are said to communicate with each other at nesting time and come ashore in groups to lay their eggs. Babies eat insects and snails and then, as they grow, plant matter, shrimp, and small fish. They take as much as fifteen years to mature, reach half a meter in length, and weigh thirty kilos.

Fly River turtles are robust animals with strong jaws, an elongated snout, and large nares, leading some to call them pig-nose turtles. These turtles use their long snouts as snorkels, enabling them to remain hidden underwater for extended periods, all the while breathing comfortably, as *Chitra indica* does, through specialized skin that allows them to extract oxygen from the water. As in elephant-nose fishes of the genus *Mormyridae* (a popular aquarium fish), Fly River turtle snouts are also electro-sensory organs, allowing them to find food in murky water.

Like leatherbacks and softshells, Fly River turtle shells are covered with leathery skin rather than the usual horny scales, but unlike softshells the bones of the carapace are domed and the bridge between the upper and lower shells is strong, making them less flat, less flexible, and a lot harder to bite than a softshell. They are fast in the water due to their hydrodynamic shape and because they use the highly webbed forelimbs to stroke in synchrony like sea turtles rather than dogpaddling like other freshwater species. They are gray to olive green overall, but with white on the lower margins of their shells and on their limbs. This white makes them harder to see from below, blending them with clouds in the sky. This is a common camouflage tactic for aquatic species and is seen in many fish. Bite-resistance, speed, maneuverability, and camouflage are all critical to their survival because these turtles share watercourses with saltwater crocodiles, some of the fiercest predators on Earth.

Like many other turtles, Fly River turtles are much in demand, as pets and food in Hong Kong and elsewhere in China. The removal of tens of thousands of these animals decimates populations, as do environmental factors, including a shrinking habitat due to gold and uranium mining operations in northern Australia. Feral buffalo destroy their nests and foul their water, and native fisherman in New Guinea use buffalo meat as bait to trap and catch them. These are an important food for aboriginal people, who represent them on rock paintings. Before the Fly River turtle was even known to Western science, it was an aboriginal totem. In 2017, the IUCN listed this species as endangered.

Giant Asian Pond Turtle

As I write this, a pair of giant Asian pond turtles (*Heosemys grandis*) stare up at me expectantly. I've had them awhile and can't remember more engaging turtles. Their appetites are insatiable, and they are as alert as any turtle I've met. Native to Myanmar, Cambodia, Thailand, Vietnam, Laos, and Malaysia, they've been thoroughly hunted in the wild. Thankfully, they are no longer commonly exported as pets, although many of them left Vietnam in the 1990s, bound for fanciers in the West and kitchens in China, where they remain prized for their size and presumably their taste. They are commonly kept in Southeast Asian temples where they can be seen piled up on land or swimming in dirty ponds. Well-intended worshippers often feed them unhealthy foods such as cupcakes.

There are much larger turtles in Asia—the giant softshell previously detailed is one example—but all are strictly aquatic. Giant Asian pond turtles, by contrast, are equally at home on water or land, though they never stray far from a lake, stream, river, creek, or pond. These turtles have massive oval brown shells with serrated margins that disappear with age. The plastron bears an attractive starburst pattern and, in young animals, a "fontanelle" or soft spot in the middle. It's an unusual feature in such a robust turtle—in any large turtle that I can think of, for that matter—and it reminds me of the soft spot on the head of human newborns, and on some Chihuahua puppies, too. With age in all the aforementioned species, bones grow together until the soft spot closes, though some very tiny teacup dogs retain it. The legs are thick and powerful, and the claws are sharp, so much so that when my young ones struggle to be free of my grasp, they dig quite painfully into my hands.

This species gives the overall impression of being the turtle version of a good-looking body builder, with a broad, handsome head and bulging jaw muscles to match. This jaw is well suited to the turtles' omnivorous diet, which includes carrion, worms, snail, bugs, and amphibians. After having kept turtles for more than half a century, I have come to suspect that there may be a correlation between the apparent intelligence of turtles and the breadth of their diet. Being opportunistic enough to be able to hunt and scavenge, forage, and lie in wait for food requires an alert awareness that makes turtles a pleasure to behold and to husband. Giant Asian pond turtles lay a small number of large eggs, which incubate for the better part of three months. As of 2019, the IUCN considers this to be a threatened species.

Leatherback Sea Turtle

Though some are physically imposing animals, sea turtles are a small group—a mere eight species in all. Employed as a symbol for marine conservation, species protection, and in efforts to modernize commercial line fishing and trawling methods, sea turtles are venerated in Micronesia and Polynesia, are the darlings of conservation efforts worldwide, and are beloved by scuba divers, snorkelers, and surfers worldwide. Their characteristic, vaguely triangular shape is rendered in tropical art, familiar in aquarium exhibits, and are often seen on T-shirts, bumper and window stickers, and on surfboards and paddleboards.

The leatherback is the largest of the sea turtles and is also the largest turtle alive in the modern world. The only species in both its genus and family, *Dermochelys coriacea* is also the turtle species with the widest naturally occurring global distribution. Its anatomy makes it so unique, scientists debate its evolutionary history, agreeing only that its overall appearance is classic sea turtle, albeit with some interesting specializations. We can't be as sure of its natural history because despite studies employing transmitters and satellites to track its movements, much of the leatherback's natural history remains mystery. We do know that leatherbacks like both ocean basins and shallows, and that they prefer to nest on beaches in Suriname, Guyana, the lesser Antilles, Colombia, Australia, West Africa, and Costa Rica. Researchers have also observed that leatherbacks like to eat jellyfish (tearing them apart with conical protuberances in the mouth) and a wide range of other marine items including crabs and urchins, fish, and smaller turtles.

We also know that leatherbacks can lay as many as a thousand eggs in a season—so many because nature is beyond cruel to nesting sea turtles, with poachers, racoons, seabirds, fish, and lizards taking a stunning percent of young before they make it to breeding age. I well remember creeping up on a nesting leatherback in the company of a park ranger on a remote Central American beach. The moon was at half-staff, the surface of the sea glinted, and the incoming waves were frothy and gentle. The giant glided ashore like the natural surfer it was, watermelon-sized head leading its return to its ancestral home. Through a complex combination of magnetic orientation and star reading, sea turtles nest exclusively on the precise beach they first see, so this beautiful beast knew just where she was going. In fact, she had likely traversed thousands of miles of open ocean to reach this destination.

Exactly how large the leatherback gets is the subject of some heated scientific debate. The legendary nineteenth-century naturalist Louis Agassiz is said to have recorded huge specimens, and there is a modern documented report of one weighing nine-hundred-and-fifty kilos. These days, however, despite heavy clots of ingested plastic in their stomachs, most weigh closer four-hundred-and-fifty kilos. Their carapaces can be two meters long or even more. Leatherbacks are supremely aquadynamic animals, the product of hundreds of millions of years of an evolutionary quest for more efficiency and speed. Water flows over its barrel-chested, blue-black shell so easily and with such little resistance that the turtle veritably slips through the water, its rear legs tucked in behind, its massive flippers out front.

The giant's blue skin is dotted with pale pink, and the top of the wedge-like flattened head possesses a white mark like a horse's blaze. The bones remain highly cartilaginous and soft, suggesting the retention of embryonic characteristics into adult life. This flexibility, along with the body-wide presence of an oil full of polyunsaturated fatty acids, enable it to survive the crushing pressure of the water column when it dives to depths of one thousand meters, as well as the frigid temperatures of the deep. The leatherback's oil makes its flesh inedible and possibly toxic and may also help it avoid nitrogen poisoning—"the bends"—when it surfaces from great depths. Too, the leatherback is one of those exceedingly rare reptiles whose circulatory system and metabolism raise its body temperature above that of the surrounding environment.

The Swiss-based International Union for the Conservation of Nature (IUCN), a group of scientists and experts keeping tabs on the status of many endangered species, determined in 2019 that populations in some locations are doing better than expected while others, the one at

Terangganu in Malaysia, for example, have been utterly annihilated. There can be no doubt that ocean plastics, brain-busting navy sonar, the killing propellers of giant ships, the entangling threat of fishing lines, beach development, egg harvesting, seismic surveys, and marine pollution negatively affect leatherbacks. Captive breeding is not possible as leatherbacks do not survive in captivity. They die of impaction when they are fed anything but gigantic quantities of jellyfish and seem unable to distinguish and avoid obstacles in their path, bashing into aquarium boundaries and features when confined. Widespread and ongoing protection programs are needed if these unique and beautiful open-ocean creatures are to survive.

Leopard Tortoise

Since turtle fanciers have come to realize that the wildly popular African spurred tortoise (*Centrochelys sulcata*) reaches an unmanageable size, they have recently returned their attention to the Leopard tortoise (*Stigmochelys pardalis*). And well they should. Although not as muscular, widespread, and imposing as its larger cousin, the leopard tortoise is, at least to my eye, a far more beautiful animal. In any case, these are Africa's two great tortoises, both are popular and widely distributed, and they are even said to interbreed. When I was a boy, the leopard tortoise was easily and often available in pet shops around New York City, and always for a song. I remember having a couple and keeping them in an aquarium empty of water— always a bad idea because turtles do not understand the concept of glass and the lack of the security of a solid border stresses them—and feeding them alfalfa hay and lettuce and other vegetables. They did all right, but they didn't thrive, mostly because they needed more heat, more space, and more freedom to choose their preferred combination of temperature and light than I could give them in an apartment. Without these things, the leopard tortoise is susceptible to fungal rotting of its shell and respiratory infections that often prove fatal. In short, it does best when kept outdoors in a hot, dry climate.

As the name suggests, the leopard tortoise has a creamy background interrupted by dark spots. These are really clear in hatchlings and young, but eventually fade to a solid tan in large, mature animals. This is one of the largest of the mainland tortoises, though the Aldabra Island and Galapagos Archipelago species grow dimensionally larger. Even so, a large leopard tortoise can weigh fifty kilos. Leopard tortoises are distributed over such a large chunk of south and east Africa, and they have successfully adapted to a wide variety of habitats, from dry savannah, to semi-deserts, to mountain uplands, to an elevation of nearly three thousand meters. Because there are so many different local populations, often with subtly different coloring, they have, in the past, been divided into more than one subspecies. Genetic investigations have, for the moment, put those distinctions to rest. While this sounds like a good thing, it may not be. Biologists argue fiercely over whether genes are the final arbiter of what defines a species or subspecies (sometimes called a race) because specific populations that have learned to exploit new ecological niches are likely on the way to becoming a different species, a process current genetic testing technology cannot yet identify. Whether genetically identical or not, natural history and behavior should, and does, play a role in deciding whether given populations are identical or different.

When it's cold, the leopard tortoise will exploit either the burrows of other animals (foxes, aardvarks, jackals) or even spaces between large rocks or termite mounds, man-sized structures of hardened earth. In extreme heat, they seek the shade of trees and bushes. Surprisingly, given the general rule that turtles with domed shells don't swim well, they have been found to cross lakes and streams without difficulty. They are creatively herbivorous, which means that while they will eat dung and bones—that last presumably when they need minerals for shell deposition—they prefer mushrooms, fruit, grasses, flowering plants like thistles, succulents, and more. The leopard tortoises I've kept have always loved the New World cactus known as *Opuntia*. In turn, when young, they are preyed upon by monitor lizards, snakes, some birds, and jackals. Most predators won't bother with adult tortoises, save for a desperately bored or hungry hyena or lion.

Leopard tortoises are roaring, grunting, wheezing copulators. Females expel fluid from their cloaca to soften soil before digging nests in protected, rocky areas wherein they deposit their eggs. They lay as often as every couple of months and produce more than a couple of dozen eggs per year. Incubation takes from a bit less than a year to nearly a year and a half, depending mostly upon outside temperature; the warmer the nest, the faster the development of the hatchlings. The IUCN, as of 2019, seems not too worried about population levels, whether because of their vast distribution or because so many are being bred in farms for the pet trade.

Marginated Tortoise

I have always found *Testudo marginata*, largest of the European tortoises, to be an especially loveable beast. Perhaps it's the pugnacious way the males ram each other with their well-developed gular scutes and the active and entertaining personalities of both sexes, who follow me around their pen, taking food from my hand, always interested in what's going on. The genus *Testudo*, which connotes classical tortoise to me, has long been celebrated in Western culture, being represented on ancient coins centuries before the common era, and seen in old paintings and on frescoes, too. These tortoises, like all true members of that group, have columnar limbs, meaning they lift themselves off the ground while walking, so their plastron does not drag. The present species has an elongated shell, which in adult males flares out widely to the rear in a kind of skirt, which can make mating a bit problematic, and results in males sliding sideways off females. It is for these special scutes that the marginated tortoise is named.

Most of this book's discussions about the interaction between turtles and humans has focused on the fact that we are driving turtles to extinction. In the case of the marginated tortoise, however, there is something more upbeat to say, namely that people seem to have been involved in the wider distribution of what began as a purely Greek species, now the nominate race *Testudo marginata*. In preparing this species profile, I learned that an ancient Indo-European people named the Ligurians—for the region of northwestern Italy from which they derive their name—may have introduced the marginated tortoise to Sardinia. That population is now genetically distinct and called *Testudo marginata sarda*. Another race, *Testudo marginata cretensis,* has gone extinct, but was on the island of Crete until the last Ice Age. It may have also been transported there by people.

There is also a dwarf species on the southwestern coast of the Greek Peloponnesus. These are phenotypically the same as other marginated tortoises, meaning that they look the same, although they are smaller, with a correspondingly smaller "skirt." Genetic studies show them to be a separate species, *Testudo weissingeri*, though to me, a marginated tortoise is a marginated tortoise. While there will be a far more impressive story of human distribution of turtles in my species discussion of the red-eared slider, I find this interaction between humans and turtles interesting, if for no other reason than that it occurred in ancient times. The bottom line seems to be that humans picked up these turtles and carried them around the Mediterranean for thousands of years, causing some subspeciation in the process.

I'm interested to know why this would be. It may not be a question anyone can ever answer, barring some unforeseen archeological revelation, but perhaps it was because people simply liked them as pets, as nature and longevity totems, or as some kind of religious or spiritual symbol. It is said that raptors take baby tortoises aloft, and may drop them high on mountain peaks where they survive and grow and can be found in the company of holy men (see Bonin, Devaux, and Dupré in my source notes) who go to these places seeking solitude. It seems doubtful that people farmed or ate them, unlike what we see in Asia and Central and South America. Perhaps Ligurians and early Greeks simply found their company as calming and satisfying as modern fanciers do.

Young tortoises bear black scutes with clear yellow centers, but these fade and blend with age. Large adults weigh as much as six kilos and reaches a length of forty centimeters. The head is dark gray or black and the forelimbs are distinguished by heavy scales on the forward-facing sides. Some tortoises have a thickened point to the tip of the tail, presumably to protect it from being bitten off or snagged, but these tortoises lack that feature. Females lay six to nine large eggs per season, occasionally a few more. As with other denizens of dry climates, heavy rains and thunderstorms cue feeding, mating, egg-laying, and hatching. Mating, they circle, ram each other, and utter cries that are surprisingly loud for their size. As some other turtles do, after having dug a hole and laid their eggs, they tamp the dirt down to seal the hole with their feet. They occur at sea level but also up to altitudes of 1,600 meters, suggesting a wide temperature tolerance. They are primarily herbivorous but will take carrion from time to time.

Marginated tortoises are well protected in their native lands, perhaps because these are First-World countries with sophisticated conservation laws and enforcement. They are CITES 2 animals, so not legally traded commercially across borders, but the IUCN regards them as a species of least concern. Even so, the future of any turtle or tortoise in today's world is far from guaranteed.

Muhlenberg's Bog Turtle

Glyptemys muhlenbergii was named by Gotthilf Heinrich Ernst Mühlenberg, a self-taught botanist and Lutheran minister who also named more than 150 species of North American plants. Elusive, emblematic, secretive, and vanishingly rare, Muhlenberg's bog turtle is the poster child for American turtle exploitation, and living (for a little while longer, anyway) proof that the war on turtles is not taking place solely in the Earth's overpopulated and undereducated countries. Like its close relative, the reasonably recently reclassified North American wood turtle, *Glyptemys insculpta*, Muhlenberg's bog turtles are less aquatic than their shape suggests. Females lay three to six small, soft eggs; hatchlings are born in autumn and almost immediately gain the confidence to adventure overland and are often not seen until the following spring. A denizen of the American Northeast, there are bog turtle populations from extreme northern Georgia, all the way up to eastern New York and western Massachusetts, although the exact locations are closely guarded secrets due to the temptation revealing them poses for poachers.

The bog turtle has a relatively round shell, reddish black with lighter colors in the centers of the scutes. There are spots on the legs and the head, which has a triangular snout and large orange blotches on each side. They're pretty heads, I think, and contribute to the overall understated elegance of the animal. I suppose that after one has been around turtles for so long, a fancier's tastes become more and more refined. I find that in my many cases, the garish colors of obviously beautiful turtles and popular turtles such as the Painted wood turtle, *Rhinoclemmys pulcherrima*, appeal somewhat less to me than more subtly rendered animals like the Painted wood turtle's relative, *Rhinoclemmys diademata,* the Maracaibo wood turtle, which like the bog turtle, has a comparatively drab shell with a colorful head.

In any case, the bog turtle is equally happy eating on land or in the water, and is an omnivore, favoring bugs and frogs and fish as well as aquatic plants, algae, and berries. It likes to be warm more than some other members of its genus and spends a lot of time basking. All the same, it can take advantage of its environment in interesting ways. I rendered it as hiding in the mud because it likes to do so. This suggests either an ability to survive without breathing for extended periods or to extract oxygen from the water as previously discussed.

Water snakes, foxes, skunks, dogs, raccoons, and snapping turtles prey upon bog turtles, but of course their biggest enemies are human beings who have brought them to them to the very precipice of extinction by collecting them in the wild. Like *Geoemyda japonica*, it is highly protected and illegal to molest, capture, or collect. My one-time boss, the late John Behler, enthusiastic naturalist and former curator of herpetology at New York's Bronx Zoo, was a great champion for saving this species, initiating protections for it as well as a breeding program. According to the IUCN in 2011, this is a critically endangered species.

Narrow-Bridged Musk Turtle

The first time I saw narrow-bridged musk turtles, *Claudius angustatus,* was at the venerable old Reptile House at New York's Bronx Zoo. I was at once smitten by their oversized heads, hooked jaws, undersized cruciform plastrons, and preposterous attitude. The turtle struck me then, and still does, as the fullest, oddest flower of Dao, a perfect example of evolutionary diversity run amok. Looking like tiny snapping turtles with a keeled carapace less than sixteen centimeters long, they swim well, but like other turtles adapted to shallow, muddy water, prefer to crawl around in search of food rather than piloting their way through open water. Their courtship involves claw waving and head bobbing, as it does for many other aquatic turtles, and the male is wont to hold on tightly while he mounts the female as she can take him for quite a ride before intromission occurs. Narrow-bridged musk turtles may nest more than once in a season, depositing two to six eggs at each laying. Depending upon conditions, the eggs take between 100 to 150 days to hatch.

These musk turtles are predatory carnivores. Pick one up at your peril, as these little turtles will happily employ their long necks to reach your fingers and shed your blood. One online turtle vendor claims that the males he has for sale enjoy biting as a hobby! I've had a couple of captive-born examples for a while now. While they are aquatic, they can and do leave the water to wander around my yard. I feed them worms, small fish, and commercial pellets, but in the wild, they eat insects, crustaceans, fish, worms, snails, newts, and frogs. I can tell you that whatever they eat, it is a satisfying business to watch them devour their chow. Like snapping turtles, they become obese quickly if overfed in captivity (a turtle shows its fat in rolls around the neck and limbs), but unlike snappers, they are quite active.

Narrow-bridged musk turtles are themselves eaten by natives in their natural range (also, I would guess, by large lizards, rodents, and birds), which includes a strip of Central America and Mexico. It seems strange to me that anyone would want to tangle with such a feisty creature for the meager meat on its bones, but, of course, hunger births desperation. Back when I first met them, narrow-bridged musk turtles were utterly unobtainable. Now, hobbyists in the USA breed them and market them with names like Vampire Musk Turtle. I've read that they are farmed for food in Mexico, and sometimes released for the good of the wild population, but I have not myself visited such a farm. As of 2019, the IUCN lists this species as near threatened.

Negev Tortoise

Some of the most important aspects of the natural history of this little tortoise have to do with its interaction with humans in this particularly hotly contested piece of Planet Earth real estate. As with so many other turtles, habitat destruction, over-collecting, road fatalities, cattle grazing, warfare, and peacetime military maneuvers have more than taken their toll. Indeed, if you think of their Middle Eastern habitat from the point of view of tortoises in the area, you might wonder what they did to be born into such a hotbed of pernicious human conflict. On the other hand, I doubt that being a turtle in the Middle East is any worse than being one in China or Southeast Asia, where turtles are relentlessly hunted as this book describes. Any way you look at it, it's not a good thing to be a turtle in a world dominated by *Homo ignorantii*.

So much habitat has been affected and so many turtles have died that even the classification of this animal is somewhat problematic. Specifically, the Negev tortoise, *Testudo werneri* is so similar to the Egyptian tortoise, *Testudo kleinmanni*, that the former was long thought to be merely a race of the latter, particularly because they look nearly identical and probably have lived alongside each other for eons. Currently, the Egyptian tortoise is seen as living only in a desperately shrinking range along a coastal strip from Libya to Egypt, whereas the Negev tortoise has an even smaller, Israeli distribution, including Be'er Sheva, Be'er Mash'abbim, Dimona, and a small piece of the Sinai Peninsula.

Like my little mentor from Namibia, this is a very small tortoise whose size is likely dictated by the rigors of the desert environment. It sports a yellow shell that sometimes fades toward green and has brown seams when the tortoise is young. Limbs, head, and neck are pale yellow with green or grayish spots, and the scales on the forelimbs are large and flat, helping it to dig and disappear under the sand when threatened. The limbs are also long, helping the tortoise walk with a high step to encourage airflow and to keep the plastron high off the burning sand, though they sleep rather than walk during summer when the temperatures are very high. Females are larger than males, but the maximum length is only 130 millimeters or so, making them considerably smaller than that useful general-reference species, the North American box turtle.

There isn't much food in the desert, so these tortoises are opportunists, consuming carrion, bugs, fallen fruit, desert flowers (particularly white or yellow ones), and whatever organic debris they can find. In suitable temperatures, Negev tortoises emerges to forage at the start of the day and retire when the heat rises too high. They gather in the shade of scrubby bushes, where the sand may be cooler and even more moist. While they are not known to drink, they are immediately afield in the rare event that rain appears. It is likely that their urine is exceedingly concentrated so as to minimize fluid loss.

As of 2019, the IUCN lists both the Negev and Egyptian tortoises as critically endangered, but conservation efforts in Israel have apparently resulted in a stabilization of the Negev tortoise population. Some accounts even have it rebounding, though it is still in desperate straits. If it is to survive, real, active conservation plans need to be enacted. One big strike against the Negev tortoise's future is that it only lays a single egg per breeding season. Worse, captive breeding programs may often hybridize it with the Egyptian tortoise, making its survival as a genetically pure species even more dubious. This is one highly specialized, vanishing creature. I hope it manages to hang on.

Okinawan Leaf Turtle

If there is one species of turtle that represents the Holy Grail to turtle hobbyists, it is the highly protected and legally unobtainable

Okinawan leaf turtle, *Geoemyda japonica*, also known as the Ryukyu black-breasted leaf turtle. This turtle is seen as a Japanese national treasure and is so strictly protected by the Japanese government that even moving one off an active roadway to save its life is prohibited. Some years ago, a turtle conservation group posted video online detailing a turtle quest into the Okinawan hinterland. The film specifically showed how and where to find these rare turtles in their native habitat. What the scientists who posted the video were thinking is not clear, perhaps they simply wanted to engage their audience and let people see and enjoy them these rare creatures without traveling to the southern islands of Japan. The sad result, however, was that in addition to entertaining Internet viewers, the film revealed how smugglers might find this elusive animal. Smugglers took the cue, resulting in a drop in the wild population.

Geoemyda was once a "catch-basin" genus into which many aquatic and semi-terrestrial species found around the world were placed. Now, with modern classification methods in place, these diverse animals occupy eight different genera. Only three of the original animals classified together remain within the formerly broadly populated genus. All from tropical Asia, these are *Geoemyda japonica*, *Geoemyda spengleri*, and *Geoemyda silvatica*. The last one, inhabiting dark, dense forests in southwestern India, is so rare, only about one hundred specimens have ever been found. Throughout its range, which includes not only Okinawa but other islands in the Ryukyu chain, the Okinawan leaf turtle also prefers the dim, filtered light of a tropical forest.

Okinawan leaf turtles are cryptically colored, meaning they blend in well with their surroundings, which are also mostly dimly lit. These small to medium-sized turtles have keeled yellow to orange shells that are usually dull, but in exceptional individuals can be brightly colored. In the familiar pattern of some Asian species, the shells bear spines when the animals are young, but become progressively smoother on the way to a maximum of fifteen centimeters. The limbs are dark with yellow or red lines and circles, and some of these lines extend to the head. The eyes are oversize and bright, lending an alert mien, and the odd tongue is heart shaped.

People who keep them in captivity—and there can be little excuse for doing so outside an institution in the case of such a rare and strictly protected species—report that they are active and engaging. Okinawan leaf turtles lays fewer than half a dozen eggs during the hot, wet summer. During the winter, they are largely dormant, hiding deep within rocky crevices, but can climb nearly steep grades and forage for earthworms, insects, roots, shoots, and mushrooms when warm weather returns. In captivity, they eat dead mice, suggesting they are opportunistic feeders. The IUCN, as of 2019, considers this species endangered.

Red-Eared Slider

There is no more ubiquitous and recognizable turtle in all the world than the red-eared slider, *Trachemys scripta elegans*. This is the proverbial pet-shop turtle, the carnival prize kids used to get—and maybe still do—for popping balloons with darts or shooting passing disks with a BB gun. It's a strange coincidence, I think, that dead-eye shooting should so often be rewarded by a prize that happens, as mentioned in my introduction, to be a creature that possesses the most complex color vision of any earthly vertebrate.

The genus *Trachemys* is a relatively recent scientific creation. When I was a boy, it belonged to another genus, and more recently, a genus different from that. All those genera were within the family *Emydidae*, along with some of the other semiaquatic turtles in this book. The red-eared slider originates in the Southeastern United States and is one of three subspecies of *Trachemys scripta*. Its close relatives occur naturally in Central and South America, in Mexico, and in Canada. It is a common turtle, not only in the wild, but in large Louisiana farms where it is bred in huge numbers for export around the world. So-called "morphs," genetic variants of these turtles, are prized by hobbyists, particularly albino versions with white shells and red eyes. In its natural form, though, it is the familiar, flattened, green aquatic turtle with strips on the shell and head, and the red marks by the ears that inspire its common name.

While they are well-distributed in American waterways from north to south, east to west, and are available from pet shops, specialty reptile shops, and general stores, pet red-eared sliders have also been introduced to Australia, Israel, the Caribbean, Guam, South Africa, multiple locations in Southeast Asia, and Great Britain. Their import has actually been banned by the European Union, not on the grounds of conservation, but out of a desire to protect indigenous wildlife. In all the places it has invaded, it is regarded as among the very worst of all invaders due to its ability to outcompete and displace native animals and wreak general environmental havoc. To cap off the picture, Red-eared sliders are also known to carry *salmonella*, bacteria which they themselves are not susceptible to, but which they can pass on to their keepers, especially to children, who have been hospitalized with the intestinal consequences of *salmonella* infection.

The red-eared slider is so successful because it is a generalist. It is able to tolerate a wide range of temperatures, is resistant to poor water quality, and is omnivorous enough to be able to make a meal of nearly anything it can bite, from pet-store pellets to carrion, bugs, plants, and garbage. It is alert, intelligent, and an inveterate basker, looking always to soak up the rays of the sun on half-submerged logs, riverbanks, and rocks. Close your eyes, conjure the image of an aquatic turtle, and it is the red-eared slider that will most likely come to mind. These are medium-sized turtles, growing to no more than twenty-eight centimeters. The male's plastron is slightly concave to accommodate the swell of the female's shell when mating, and males are a bit smaller. Males have longer claws than females do, which help them hold onto the female's shell during copulation. They are fecund creatures, laying up to thirty eggs as many as five times per year.

I cannot think of another species of turtle whose fate is so *successfully* intertwined with human beings. While some ancient people may have carried a few marginated tortoises around, red-eared sliders have exploded across the globe as instruments of human commerce. Think of them as reptilian smart phones, perhaps, or the turtle version of Starbucks coffee. I find it no coincidence that in both humans and red-eared sliders, nature favors the nervous system, anatomy, physiology, and evolutionary history of the generalist so as to exploit the maximum possible number of habitats. If, in the future, there is some global cataclysm that causes the extinction of humans and many other of the world's creatures, my money is on the red-eared slider to be one of the species that survives.

Perhaps future sliders will pen fables to share with their children in which long-extinct animals called human beings fooled with Mother Nature and suffered the consequences.

Stupendemys

Eons after Planet Earth was a smoldering rock covered by a stew of organic molecules just waiting to form life, vertebrates first crawled out on land and laid hard-shelled eggs. Some of those vertebrates become proto-turtles, and those proto-turtles became what some paleontologists (turtle classification is a surprisingly controversial scientific topic) call the ancient clade, *Archosauromorpha*. In this group was a dizzying array of shapes and colors, sizes, adaptations, and specializations. One of the most obvious specializations remains the way in which turtles retract their heads. A good number pull the head straight back in; others pull their heads in sideways. The first comprise the group *Cryptodira*, the second the *Pleurodira*.

Stupendemys, possibly the largest turtle ever to have lived, was a pleurodire with a weird, collar-like upturn of the carapace behind its head. That carapace was more than three meters long, and as such was even larger than *Archelon*, the ancient marine giant mentioned as an ancestor by the leatherback in this book's first fable. For contrast, the largest pleurodire living now, the giant Amazon River turtle, *Podocnemis expansa*, is barely a third as long, though a much better swimmer than *Stupendemys* ever was. Indeed, *Stupendemys* spent most of its time walking on the bottom of shallow watercourses. What it ate is anyone's guess because we don't know all the things that grew, swam, crawled, and slithered back then, but extrapolating backward from existing turtles, we can guess it was probably an opportunistic omnivore, favoring whatever prey items it could find, and vegetation, too.

Its South American home would be unrecognizable to us, for during the Miocene and early Pliocene periods (between twenty-three and two and a half million years ago) shallow seas covered much of what is now dry land and dry land existed where some seas are now. *Stupendemys'* specific habitat was a low-lying rainforest mixed with swamps and flood plains, an environment it shared with catfish, lungfish, sharks, rays, river dolphins, early monkeys, giant ground sloths, huge rodents, and a small, predatory bat whose Latin name, *Noctilio lacrimaelunaris,* means "tears of the moon."

Researching this ancient giant, I came across the fact that *Stupendemys* is the basis for a fictitious, crystal-shelled turtle character in an online video game. I don't know what this means other than that folks besides this Daoist monk have found *Stupendemys* an inspiration. I wonder if gamers, dreaming about this prehistoric giant, have credited it with the wisdom it showed me in my fable.

Vietnamese Box Turtle

There is no more endangered group of turtles in the world than the turtles of Asia. Within that unfortunate assemblage, none are more critically imperiled than the box turtles belonging to the genus *Cuora*, within the family *Geoemydidae*. Because modern classification methods are taking an ever more balanced approach to both morphology (form) and genetics (DNA), the composition of both family and genus are in gentle flux, with animals moving in and of the groups as more studies are published. In any case, *Cuora* are small to medium-sized turtles that can often close their shells to predation like an American box turtle, and, for the most part, spend a good deal of time both on land and in the water. Anyone familiar with the various species of *Cuora* might take their dire predicament to be a direct result of their beauty and outgoing personality—and surely those are contributing factors. The real reason for their decline, though, is a combination of the development and pollution that arise from unbridled population growth in East and Southeast Asia and the special place given to these turtles in traditional Chinese medicine and culture.

The animal I met during meditation was the Vietnamese three-striped box turtle, *Cuora cyclornata*. This species was for a long time thought to be identical to the so-called golden coin turtle, *Cuora trifasciata*, but genetic markers, a larger size, and a different geographic range differentiate the two. That larger size is to the Vietnamese animal's great misfortune, for the Chinese prize larger turtles, primarily because they yield meat for consumption and more shell for medicine. That first utility is obvious: people like to eat them. The second utility has as much to do with the current "dark economy" in China as it does with a traditional veneration of the turtle, symbol of the Earth Mother. Simply put, since personal and financial privacy has become all but impossible in their country, Chinese people have developed unique ways to circumvent government scrutiny. One of these is to do the equivalent of money laundering using turtles. Many of the most popular kinds of turtles are farmed in China and are the equivalent of commodity funds based upon their prices, which rise and fall according to supply and demand. Individual turtles, too, can be convenient, valuable, coveted, and untraceable gifts, used to reward employees or to honor friends.

As for the third utility, the plastron of these animals is used in the manufacture of something called turtle jelly—*guilinggao* in Mandarin. When combined with Chinese herbs (honeysuckle flower, rehmannia root, chrysanthemum, dandelion, forsythia, smilax rhizome, and more) along with sugar or honey, the powdered shell of *Cuora* becomes both a popular dessert and a medicine controversially prized for supporting kidney function, improving skin complexion, quelling insomnia, improving muscularity, and even for treating cancer. Despite harboring an abiding passion for traditional Chinese culture, a great respect for traditional Chinese medicine, and the belief that any medical system that has been tested *in vivo* by a large population over a great span of time and found to be effective should not be taken lightly, I am not aware of any scientific substantiation of these claims. Should they be found to be true, I would hope that other less destructive products are developed. Generally speaking, I bridle at the idea that animals should be seen as useful to humanity as opposed to sentient beings with their own inherent value and right to exist.

Despite all the farming, as of 2019, the IUCN considers nearly all *Cuora* critically endangered. In an effort to protect and propagate the future of the genus *Cuora*, I do keep a handful of captive-bred examples of these charming and beautiful creatures in my home. It is true that they live a life far from their homeland—far even from any jungle stream—but they are greatly pampered by a doting monk who loves them and has joined efforts to assure they will persist long after their native environment has been destroyed.

MY LIFE LIST

Birders keep what they call life lists. These serve as a combination of sport, personal journal, and means of remembering special moments spent in nature. They also give birders a theme for wildlife-oriented travel, build social bridges, and inspire conservation. Too, they are a measure of how well a person with a passion has managed to engage and indulge that passion over the course of a life. Such lists have a bit of a consumptive quality, as to see a bird and put it down on a list is a bit like eating it, minus the killing and the calories. However snarky and competitive birding may occasionally be, overall, I think it's a wonderful pastime. After all, the very folks who are out appreciating birds and nature might otherwise be bowling or playing first-person shooter games online or binge-watching television shows.

I'm not aware that turtle fanciers have lists. Perhaps this is because turtles don't announce themselves with calls or flap noisily overhead. If the avian palette is spacious and grand, the turtle life palette is intimate and small. Since turtles are among the most endangered of all vertebrates and are spread over most of vanishing temperate and tropical parts of the world, actually finding them is expensive, difficult, and frustrating for most people. All the same, people *keep* turtles just as they keep parrots and mynahs and finches and pigeons. While many pet turtles have been bred in captivity, even more are still removed from the wild, a practice that is not only disruptive and cruel, but which existentially threatens the very animals we purport to cherish just as the Sixth Great Extinction, caused by human activity, unfolds around us.

Why do we keep turtles? A few reasons. First, to satisfy our desire—born of our incarceration in artificial environments such as offices, houses, airports, cars, factories—to reconnect with wildness. Remembering that we are not alone but part of something larger feeds us spiritually and staves off anxiety and depression, alienation, and loneliness. Our nervous system, after all, is evolutionarily calibrated to respond to wildlife and nature. Second, turtles are flat-out *interesting* creatures. Their presence, in all their sizes, shapes, colors, and behaviors, lends an air of exoticity to our lives and can be a source of ongoing entertainment and fascination. Third, rare and expensive turtles can become lust objects for collectors. Fourth, some people have a strong need to nurture and care for other living creatures and may be allergic to animals with feathers or hair and find a turtle fits the bill. Last is the opportunity to contribute to the conservation of endangered species by purchasing animals bred in captivity and breeding them in turn.

While animal rights activists might legitimately argue against the presumed suffering of captive individuals, at this late stage in the global extinction process, captive breeding can keep endangered turtles alive long enough to protect and restore their habitats. It's not a perfect solution, but it is a case of the good of the few surrendering to the good of the many. Too, most hobbyists who are dedicated enough to breed endangered species are very concerned with the individual's welfare and happiness, if only because it takes a well, happy turtle to produce young. Zoos, which used to be carnivalesque horror shows, are now becoming important repositories of genetic material and the last, best hope for many types of turtles.

Having said all that, I cannot say that the number of animals I have produced and sold during the decades of my life with turtles comes anywhere close to the number of animals I have bought, and which have died in my care, either because of the trauma of transportation, old age, or my own ineptitude. I am far from alone in this, and painfully recognize I am more part of the problem than part of the solution. These days, I husband very few turtles, and all are members of species so massively under attack in the wild that my modest attempts to breed and preserve them are, frankly, of no consequence at all.

Here then, is my dirty laundry, the list of turtles I have either worked with or kept myself. As a turtle lover, it is very slightly a mark of my experience, expertise, and savvy, and far more significantly a mark of my great folly. I share it in closing a book about the plight of the world's turtles and in the likely fruitless hope of redemption for my admittedly minor role in their demise.

1. North American Eastern Mud Turtle *Kinosternon subrubrum subrubrum*

2. North American Yellow Mud Turtle *Kinosternon flavescens*

3. Red-Cheeked Mud Turtle *Kinosternon scorpioides cruentatum*

4. Tabasco Mud Turtle *Kinosternon acutum* (near threatened)

5. Oaxacan Mud Turtle *Kinosternon oaxacae*

6. Mexican Narrow-Bridged Mud Turtle *Claudius angustatus* (near threatened)

7. Central American River Turtle *Dermatemys mawii* (critically endangered)

8. North American Common Musk Turtle or "stinkpot" *Sternotherus odoratus*

9. North American Southern Musk Turtle *Sternotherus minor minor*

10. Central American Giant Musk Turtle *Staurotypus triporcatus* (near threatened)

11. Pacific Coast Giant Musk Turtle *Staurotypus salvinii* (near threatened)

12. North American Snapping Turtle *Chelydra serpentina*

13. North American Alligator Snapping Turtle *Macrochelys temminckii* (vulnerable and legally protected)

14. North American Eastern Painted Turtle *Chrysemys picta picta*

15. North American Western Painted Turtle *Chrysemys picta bellii*

16. Mississippi Map Turtle *Graptemys pseudogeographica kohni*

17. Barbour's Map Turtle *Graptemys barbouri* (vulnerable)

18. North American Ringed Sawback Turtle *Graptemys oculifera* (vulnerable)

19. North American Black-Knobbed Sawback Turtle *Graptemys nigrinoda*

20. North American Gulf Coast Box Turtle *Terrepene carolina major*

21. North American Ornate Box Turtle *Terrepene ornata*

22. Florida Box Turtle *Terrepene carolina bauri*

23. North American Common Box Turtle *Terrepene carolina carolina*

24. North American Three-Toed Box Turtle *Terrepene carolina triunguis*

25. North American Diamondback Terrapin *Malaclemys terrapin* (near threatened)

26. Florida Chicken Turtle *Deirochelys reticularia chrysea*

27. Florida Cooter *Trachemys floridana*

28. North American River Cooter *Trachemys concinna*

29. North American Yellow-Bellied Turtle *Trachemys scripta scripta*

30. North American Spotted Turtle *Clemmys guttata* (endangered)

31. North American Wood Turtle *Glyptemys insculpta* (endangered)

32. Pacific Coast Pond Turtle *Actinemys marmorata*

33. Colombian Slider *Trachemys callirostris*

34. Southeast Asian Leaf Turtle *Cyclemys dentata* (near threatened)

35. Mediterranean Pond Turtle *Mauremys leprosa*

36. Chinese Bigheaded Turtle *Platysternon megacephalum* (endangered)

37. Central American Painted Wood Turtle *Rhinoclemmys pulcherrima manni*

38. Central American Furrowed Wood Turtle *Rhinoclemmys areolata* (near threatened)

39. Giant Asian Pond Turtle *Heosemys grandis* (vulnerable)

40. Malaysian Box Turtle *Cuora amboinensis* (vulnerable)

41. Taiwanese Yellow-Margined Box Turtle *Cuora flavomarginata* (endangered)

42. Hainan Island Three-Striped Box Turtle *Cuora trifasciata luteocephala* (critically endangered)

43. Vietnamese Box Turtle *Cuora cyclornata cyclornata* (critically endangered)

44. Thai and Lao Box Turtle *Cuora cyclornata annamitica* (critically endangered)

45. McCord's Box Turtle *Cuora mccordi* (likely extinct in the wild)

46. Keeled Box Turtle *Cuora mouhotii* (vulnerable)

47. Chinese Pond Turtle *Mauremys reevesii* (endangered)

48. Indian Roofed Turtle *Pangshura tecta*

49. Asian Brown Roofed Turtle *Pangshura smithii* (near threatened)

50. Asian Black Marsh Turtle *Siebenrockiella crassicola* (vulnerable)

51. Asian Crowned River Turtle *Hardella thurjii* (vulnerable)

52. Mekong River Snail-Eating Turtle *Malayemys subtrijuga* (vulnerable)

53. South Asian Spotted Pond Turtle *Geoclemys hamiltonii* (vulnerable)

54. Greek Tortoise *Testudo graeca* (vulnerable)

55. African Spurred Tortoise *Centrochelys sulcata* (vulnerable)

56. Hermann's Tortoise *Testudo hermanni* (near threatened)

57. Indian Star Tortoise *Geochelone elegans* (vulnerable)

58. Burmese Star Tortoise *Geochelone platynota* (critically endangered)

59. Madagascan Radiated Tortoise *Astrochelys radiata* (critically endangered)

60. Amazon Yellow-Foot Tortoise *Chelonoidis denticulata* (vulnerable)

61. Galapagos Giant Tortoise *Chelonoidis* sp. (critically endangered)

62. Red-Foot Tortoise *Chelonoidis carbonarius*

63. South American Chaco Tortoise *Chelonoidis chilensis* (vulnerable)

64. Aldabra Giant Tortoise *Aldabrachelys gigantea* (vulnerable)

65. Burmese Brown Tortoise *Manouria emys emys* (endangered)

66. Burmese Black Mountain Tortoise *Manouria emys phayrei* (endangered)

67. Indian Elongated Tortoise *Indotestudo elongata* (endangered)

68. Celebes Tortoise *Indotestudo forsteni* (endangered)

69. Serrated Hinge-Back Tortoise *Kinixys erosa*

70. Bell's Hinge-Back Tortoise *Kinixys belliana*

71. Home's Hinge-Back Tortoise *Kinixys homeana* (vulnerable)

72. Speke's Hinge-Back Tortoise *Kinixys spekii*

73. Travancore Tortoise *Indotestudo travancorica* (vulnerable)

74. Texas Tortoise *Gopherus berlandieri*

75. North American Desert Tortoise *Gopherus agassizii*

76. North American Gopher Tortoise *Gopherus polyphemus* (vulnerable)

77. South African Bowsprit Tortoise *Chersina angulata*

78. European Marginated Tortoise *Testudo marginata*

79. South African Geometric Tortoise *Psammobates geometricus* (critically endangered)

80. South African Tent Tortoise *Psammobates tentorius*

81. Madagascan Spider Tortoise *Pyxis arachnoides* (critically endangered)

82. Southeast African Leopard Tortoise *Stigmochelys pardalis sp.*

83. South African Karoo Dwarf Tortoise *Chersobius boulengeri* (near threatened)

84. Pancake Tortoise *Malacochersus tornieri* (vulnerable)

85. Central Asian Tortoise *Testudo horsfieldii* (vulnerable)

86. Kemp's Ridley Sea Turtle *Lepidochelys kempii* (critically endangered)

87. Florida Soft-Shelled Turtle *Apalone ferox*

88. Yellow-Spotted Amazon Turtle *Podocnemis unifilis* (vulnerable)

89. Red-Headed Amazon Turtle *Podocnemis erythrocephala* (vulnerable)

90. Indian Spotted Soft-Shelled Turtle *Lissemys punctata*

91. South American River Turtle *Podocnemis expansa*

92. African Black-Bellied Hinged Terrapin *Pelusios subniger*

93. Okavango Mud Turtle *Pelusios bechuanicus*

94. Adanson's Mud Turtle *Pelusios adansonii*

95. Australian Snake-Necked Turtle *Chelodina longicollis*

96. South American Matamata Turtle *Chelus fimbriatus*

97. South American Twist-Neck Turtle *Platemys platycephala*

98. New Guinea Fly River Turtle *Carettochelys insculpta* (vulnerable)

99. African Helmeted Terrapin *Pelomedusa subrufa*

100. Okinawan Leaf Turtle *Geoemyda japonica* (endangered)

101. Vietnamese Black-Breasted Leaf Turtle *Geoemyda spengleri* (endangered)

102. Chinese Stripe-Necked Turtle *Ocadia sinensis* (endangered)

103. Asian Spiny Turtle *Heosemys spinosa* (endangered)

104. Giant Asian Pond Turtle *Heosemys grandis* (vulnerable)

105. Yellow-Headed Temple Turtle *Heosemys annandalii* (endangered)

106. Vietnamese Pond Turtle *Mauremys annamensis* (near extinction in the wild)

107. Caspian Terrapin *Mauremys caspica*

108. Southeast Asian Four-Eyed Turtle *Sacalia quadriocellata* (endangered)

109. North American Red-Eared Slider *Trachemys scripta elegans*

A NOTE ON SOURCES, TRANSLITERATIONS, AND NUMBERS

Until recently, there has been a great deal of confusion around the transliteration of Chinese words. So much so, that there were numerous systems in place to make Chinese, a tonal language, comprehensible to the English-language speaker. None of these systems has been perfect. The best may be the one developed at my alma mater, Yale University, but unfortunately few people use it. Most popular until a few years ago was the Wade-Giles system, which gave us familiar spellings such as Lao Tzu, Sun Tzu, Daoism, and more. All other systems have now largely been abandoned in deference to *Pinyin*, the standard transliteration method advanced by the Chinese Government. Following this system, we get Daoist not Taoist, Laozi not Lao Tzu, Sunzi not Sun Tzu, and so on.

Also, since I relied upon scientific sources for the measurements of turtles in this book, I have adhered to the scientific standard, which is now metric. Therefore, meters not feet, kilometers not miles, and so on. I recognize that this may be a slight inconvenience for some readers, but at-a-glance conversion sites are available online, and I wanted to maintain a modern, global, and scientific feel in the sections of this book where numbers are employed.

The vast majority of the material in this book derives from my own half century of direct experience with turtles of all shapes and sizes and kinds. In searching for scientific details, I did what everyone seems to do these days and used the Internet for images, wikis, and other readily searched and available sources. I even watched a few YouTube videos of turtles (bog turtles eating, for example) when I had either never met a turtle in person or had not seen one in decades.

In addition, I drew heavily upon the contents of my own library, which while overwhelmingly possessed of Daoist texts, does contain a few works about turtles. Notable among these is zoology legend Peter Charles Howard Pritchard's classic *Living Turtles of the World*, the bible of my youth. I acquired my personal copy of this now out of date work back in the 1970s, and as kids do, I scribbled all kinds of notes in the margins of it, defacing that seminal work, which was subsequently updated as *Encyclopedia of Turtles* (1979, Neptune, New Jersey, T.F.H. Publications, Inc.).

Professor Pritchard passed away as I was proofing these pages. The information in his books is now decidedly out of date, not for any lack of scholarship on the author's part, but because of the rapidly changing and degrading situation turtles face in the world and because of the efforts of devoted field biologists during the decades since his research was done. Professor Pritchard himself translated into English a French tome written by three of his friends and one upon which I relied extensively. This is *Turtles of the World* (Franck Bonin, Bernard Devaux, Alain Dupré, 2006, Baltimore, Maryland, The Johns Hopkins University Press). Many turtle fanciers I know cherish this book and rely upon it as the go-to source for quick information about any turtle species. It is well written, readably organized, and contains clear photographs of every species, though not so very many as some readers might wish.

More recently set up as an intellectual journey through the biological phenomenon we call the turtle, rather than as a field guide or quick reference for information about a particular species, is *Tortoises and Terrapins: A Natural History* (Ronald Orenstein, 2012, Canada, Firefly Books). In a similar line, the lighter, less exhaustive, and more recent *Turtles: An Extraordinary Natural History 245 Million Years in the Making* (Carl J. Franklin, 2011, New York, New York, Crestline/Voyageur Press) is a nice read and a good blending of species accounts and biological characteristics of the group. *Turtles, Tortoises and Terrapins* (Fritze Jurgen Obst, translated by Sylvia Furness, 1986, Melbourne, Australia, The Macmillan Company of Australia) is a slightly older volume that also gets into evolutionary and anatomical details and includes some nice drawings. Some of the information in this last work, however, may not be as accurate as more recent works, as it was published before the genetic revolution in taxonomy.

Turtles as Hopeful Monsters (Olivier Rieppel, 2017, Bloomington, Indiana, Indiana University Press) may offer the last word on turtle evolution and classification, as well as details of the history of this line of inquiry. It is a treasure trove for the scientist, not so much for the casual fancier or reader. I also have a few issues of the journal of the International Turtle and Tortoise Society, a now defunct operation that gave voice to hobbyists and breeders worldwide back when I was a boy. These are not much available anymore except through herpetological booksellers, but they did inspire me with memories of my childhood with my shelled friends.

ACKNOWLEDGEMENTS

Nine deep bows to the community of people who share my love of turtles, whether they be academic researchers, hobbyist breeders, conservationists, institutional curators, or volunteers out protecting animals and eggs. Without your kinship, infectious enthusiasm, and staunch support over the years, I don't know how or if my own passion could have endured for so long. Thanks to Jay and Danielle Frewer, turtle purveyors *par excellence*, for their friendship and for helping me find the right shelled companions, to Chris Hansen for his sometimes inscrutable but always wise turtle counsel, to Anthony Pierlioni, for the endless hours of lively turtle talk, and to my own turtle pets who have taught me so much about nature and spirit. I'm particularly grateful to all the good folks at Mango Publishing for making this book possible, in particular the visionary Brenda Knight for supporting the unorthodox idea of it, MJ Fievre for her persistent and perspicacious edit, the creative team for making it so beautiful, and to my friends and students who suffered through my requests for proofreading and suggestions. Nine bows, too for the collaborative effort between Nancy Bautista and Janelle Rosenfeld that resulted in this book's beautiful cover. These days, more than ever, a book takes a village.

Readers interested in contributing to turtle conservation will find many worthy organizations out there. Here, in no particular order, are just a very few deserving of support. If I've left you out, it doesn't mean I don't love you!

The Chelonian Research Institute:

chelonianri.org

The Turtle Conservancy:

ww.turtleconservancy.org

The Cuora Conservation Center:

www.cuora.org

The Turtle Survival Alliance:

turtlesurvival.org

The Turtle Room:

theturtleroom.com

The Turtle and Tortoise Preservation Group:

www.ttpg.org

Sea Turtle Conservancy:

conserveturtles.org

ABOUT THE AUTHOR

Dedicated to the welfare of all sentient beings, Yun Rou (the name means Soft Cloud) has been called "the new Alan Watts" for his Daoist teachings and the "Zen Gabriel Garcia-Marquez" for his works of magical realism set in China. Born Arthur Rosenfeld in New York City, he received his academic background at Yale, Cornell, and the University of California and was officially ordained a Daoist monk in Guangzhou, China. His award-winning books, optioned for film in both Hollywood and Asia, bridge spirituality, philosophy, and history. Host of the National Public Television show *Longevity Tai Chi*, he is a forty-one-year kung fu master with students and followers around the world.

Mango Publishing, established in 2014, publishes an eclectic list of books by diverse authors—both new and established voices—on topics ranging from business, personal growth, women's empowerment, LGBTQ studies, health, and spirituality to history, popular culture, time management, decluttering, lifestyle, mental wellness, aging, and sustainable living. We were recently named 2019's #1 fastest growing independent publisher by *Publishers Weekly*. Our success is driven by our main goal, which is to publish high quality books that will entertain readers as well as make a positive difference in their lives.

Our readers are our most important resource; we value your input, suggestions, and ideas. We'd love to hear from you—after all, we are publishing books for you!

Please stay in touch with us and follow us at:

Facebook: Mango Publishing
Twitter: @MangoPublishing
Instagram: @MangoPublishing
LinkedIn: Mango Publishing
Pinterest: Mango Publishing

Sign up for our newsletter at www.mangopublishinggroup.com and receive a free book!

Join us on Mango's journey to reinvent publishing, one book at a time.

Printed in the USA
CPSIA information can be obtained
at www.ICGtesting.com
JSHW031710140824
68134JS00038B/3624